Air Pollution and the Brain

Katja Marika Kanninen • Anthony Robert White
Editors

Air Pollution and the Brain

Springer

Editors
Katja Marika Kanninen
A.I. Virtanen Institute for Molecular Sciences
University of Eastern Finland
Kuopio, Finland

Anthony Robert White
Molecular and Cellular Neurodegeneration
QIMR Berghofer Medical Research Institute
Herston, QLD, Australia

ISBN 978-3-031-99301-5 ISBN 978-3-031-99302-2 (eBook)
https://doi.org/10.1007/978-3-031-99302-2

This Springer imprint is published by the registered company Springer Nature Switzerland AG
The registered company address is: Gewerbestrasse 11, 6330 Cham, Switzerland

Preface

While the adverse effects of air pollution on human health have been known for several decades, the focus has primarily been on pulmonary and cardiovascular toxicity. More recently, evidence has been accumulating from epidemiological and experimental studies to show that the nervous system is also a target for air pollution toxicity. Accumulating evidence shows that exposure to excessive air pollution is associated with neurodegenerative, neurodevelopmental, and mental disorders.

This book provides an up-to-date summary of how air pollution influences the brain. It discusses the types of pollutants that are most likely to impact the brain, pollutant translocation to the brain, and the cellular mechanisms induced by exposure. It also touches upon the current knowledge about air pollution links to neurodevelopmental, neurodegenerative, and mental disorders, while also accounting for climate change impacts and regulatory frameworks targeting air quality.

Kuopio, Finland — Katja Marika Kanninen
Herston, QLD, Australia — Anthony Robert White

Contents

About the Editors

Katja Marika Kanninen is Professor of Cellular Neurobiology. She leads a research group at the University of Eastern Finland in Kuopio, Finland. Her research utilizes highly translational human cell cultures and animal models that aim to provide cellular and molecular level understanding on the role of airborne agents on the brain in both health and disease.

Anthony Robert White, PhD, is Associate Professor of Cellular and Molecular Neurodegeneration at QIMR Berghofer Medical Research Institute, Herston, Queensland, Australia.

Chapter 1
Climate Change Impacts on Air Pollution and Effects on Brain Health

Anthony Robert White, Emily Rose Vivian, and Hazel Quek

Keywords Neuroinflammation · Oxidative stress · Mental health · Cognitive dysfunction · Environmental neurotoxicity · Aging brain

1.1 Introduction

1.2 Climate Change and Air Pollution

Climate change has led to significant variations in global climate patterns, resulting in a range of environmental impacts. These changes are characterized by rising temperatures, more frequent extreme weather events, and a general increase in the intensity of natural disasters [1, 2]. One of the most significant, but often underappreciated, outcomes of climate change is the effect on air quality [1]. As the Earth's climate continues to warm, the interactions between atmospheric conditions and pollutants have created complex dynamics that exacerbate air pollution globally [2–4].

1.2.1 Key Mechanisms Linking Climate Change and Air Pollution

Climate change influences air pollution in several ways. Higher global temperatures and heatwaves can lead to increased ground-level ozone (O_3) formation [3]. Ozone can be formed when nitrogen oxides (NO_x) and volatile organic compounds

A. R. White (✉) · E. R. Vivian · H. Quek
Brain and Mental Health Program, QIMR Berghofer Medical Research Institute, Herston, QLD, Australia
e-mail: tony.white@qimrberghofer.edu.au

K. M. Kanninen, A. R. White (eds.), *Air Pollution and the Brain*,
https://doi.org/10.1007/978-3-031-99302-2_1

(VOCs), released from vehicles and industrial activities, react in the presence of sunlight [5]. With rising temperatures, the rate of these photochemical reactions increases, leading to higher O_3 concentrations, especially in urban/industrial environments [1, 3, 6].

In addition, climate change contributes to more frequent wildfires, significantly increasing the concentration of particulate matter (PM) and additional pollutants in the atmosphere. Wildfires release large amounts of fine PM (PM2.5 microns), which can be carried over long distances, affecting air quality in regions distant from the fire source [7]. The PM released by such events contains a mixture of organic and inorganic materials, which can be toxic when inhaled and have been associated with a range of adverse health effects [8]. Changes in atmospheric circulation also play an important role in air pollution. Alterations in wind patterns can trap pollutants in certain areas, preventing their dispersal and leading to the stagnation of polluted air over cities and regions [9]. This phenomenon is particularly concerning in areas with high levels of industrial activity, vehicular emissions, or woodfire smoke generation, leading to hazardous conditions for human health.

1.2.2 Global Trends in Air Pollution and Climate Change

While climate change amplifies air pollution across the globe, the effects are not uniform. Urban areas, particularly those in developing countries, are experiencing higher concentrations of pollutants due to both local emissions and the compounded effects of climate-related factors [1, 10]. Developing nations often experience rapid industrialization and population growth, which further exacerbates air quality issues. Coupled with weaker environmental regulations and the increasing frequency of climate-driven events, these regions face disproportionately higher levels of pollution-related health risks [11]. Rural areas can also be greatly affected by air pollution. In addition to biomass burning, dust storms, exacerbated by desertification and changing weather patterns, contribute to poor air quality in regions including the Middle East and northern Africa [12, 13]. Similarly, the melting of Arctic ice is releasing trapped pollutants such as mercury and arsenic into the atmosphere, creating new challenges for air quality management in the northern hemisphere [14].

1.2.3 Human Health and Environmental Implications

The convergence of climate change and air pollution presents a serious public health challenge. Air pollution, already a leading cause of premature mortality globally, is becoming more intense and widespread due to climate-related factors. PM2.5 and O_3, in particular, have been linked to respiratory, cardiovascular, and increasingly, brain diseases and effects on cognitive function [1–4]. The health impacts of climate change are diverse and widespread, affecting multiple systems. Heat-related illnesses are becoming more common due to frequent and intense heat waves, leading to conditions such as heat exhaustion and heatstroke, particularly in vulnerable populations such as the elderly and children [15]. Urban areas experience heightened risks due to the "urban heat island" effect. Climate change is also reshaping the transmission of infectious diseases, with warmer temperatures expanding the habitats of disease vectors, i.e., mosquitoes, leading to the spread of malaria and dengue to new regions [16]. Finally, food and water security are threatened by changing weather patterns, affecting agricultural yields and leading to food shortages and malnutrition, especially in low-income areas [17].

1.2.4 Mental Health Impacts

The mental health impacts of climate change are becoming increasingly recognized. The psychological toll of extreme weather events can lead to anxiety, depression, and post-traumatic stress disorder (PTSD) [18]. The uncertainty and stress associated with the slow-onset impacts of climate change also contribute to chronic stress and anxiety [18]. In addition, the mental health of children and adolescents is particularly vulnerable, as they face an uncertain future shaped by climate change, which can lead to eco-anxiety, feelings of helplessness, and other emotional responses [19].

1.3 Brain Health and Climate Change

Understanding how climate change impacts brain health has emerged as a critical area of research due to the growing recognition that the environment plays a fundamental role in neurological development, functioning, and the risk of disease [20] (Fig. 1.1). While genetic factors are often associated with neurological disorders, mounting evidence suggests that environmental exposures, ranging from air pollution to chemical toxins, heat stress, and noise pollution, are significant contributors to brain health outcomes [21] (Fig. 1.1). Investigating the relationship between these environmental factors and brain health is crucial for identifying preventive strategies and mitigating the rising burden of neurological diseases worldwide [8].

Air pollution is increasingly recognized as a major contributor to neurological disorders. PM2.5, NOx, and carbon monoxide (CO) can bypass the body's natural defenses and penetrate the brain, causing oxidative stress and inflammation [22]. This has been linked to cognitive decline, strokes, and an elevated risk of neurodevelopmental disorders such as autism spectrum disorder (ASD) and attention deficit hyperactivity disorder (ADHD) [23]. In adults, long-term exposure to pollutants has also been associated with Alzheimer's and Parkinson's diseases. With air pollution affecting millions globally, its impact on brain health requires urgent attention [22].

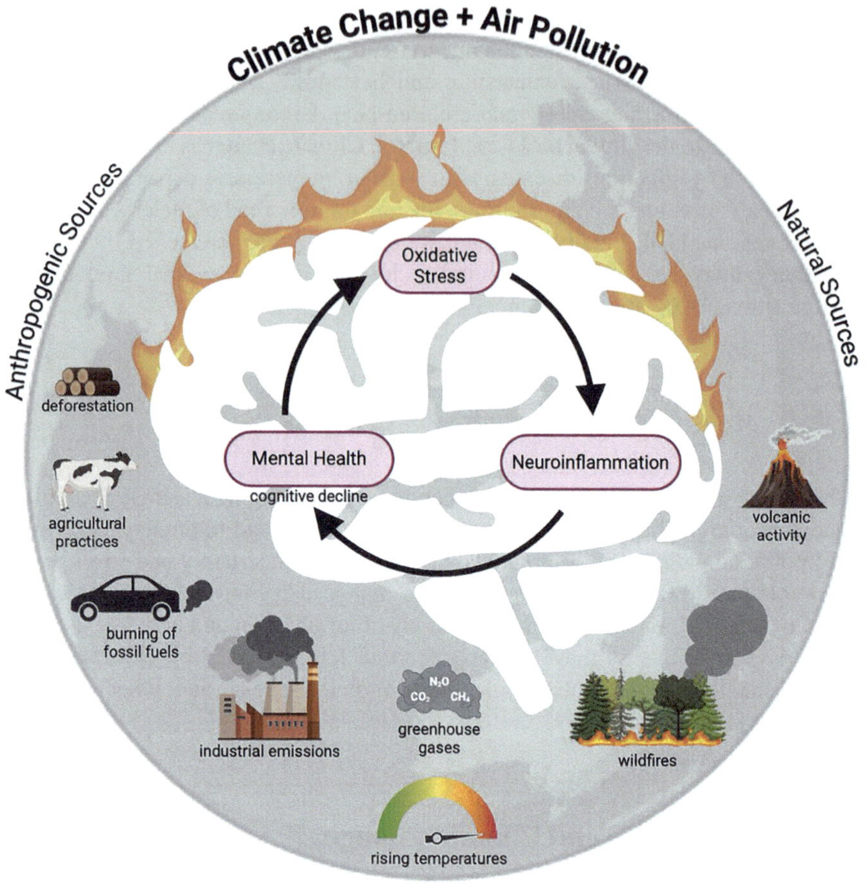

Fig. 1.1 Impact of climate change and air Pollution on brain health. This schematic illustrates the interconnected pathways through which climate change and air pollution-originating from both anthropogenic (left) and natural (right) sources—contribute to adverse neurological outcomes. Anthropogenic sources include deforestation, agricultural practices, burning of fossil fuels, industrial emissions, and the release of greenhouse gases (CO_2, CH_4, N_2O), all of which contribute to rising global temperatures. Natural sources such as volcanic activity and wildfires also exacerbate air pollution and climate instability. These environmental stressors contribute to increased oxidative stress and neuroinflammation in the brain, which in turn can impair mental health and accelerate cognitive decline. This cycle reinforces itself, creating a feedback loop that links environmental exposures to long-term neurological and psychological consequences. Overall, highlighting the urgency of addressing environmental health challenges to mitigate risks to brain health across populations

Climate change further exacerbates these issues. Frequent heat waves increase stroke (ischemia) rates and trigger neurological conditions, as prolonged heat exposure can disrupt cognitive functioning [24]. In addition, events such as wildfires and hurricanes cause acute and chronic stress, increasing the prevalence of mental health disorders such as anxiety, depression, and PTSD [25]. Vulnerable populations, including the elderly, are particularly at risk from climate-induced neurological stress [8]. Beyond climate factors, exposure to airborne neurotoxic chemicals, including heavy metals and industrial pollutants, poses another significant risk. Toxic molecules, including lead, arsenic, mercury, and pesticides, have been shown to impair neurodevelopment in children, reduce cognitive function, and heighten the risk of neurodegenerative diseases [26]. Developmental exposure to environmental hazards represents a critical concern. Prenatal and early childhood exposure to air pollutants can have lasting effects on neurological health [27]. Air pollution during pregnancy has been linked to reduced brain volume and cognitive deficits in children, while exposure to toxic chemicals can impair learning and memory [28]. Identifying critical windows of vulnerability is essential for developing protective measures and policies. Given the rising global burden of neurological disorders and the worsening impact of climate change and air pollution, research into the complex links between environmental factors and brain health is critically important. This research is crucial for informing public health priorities and promoting safer environments, particularly for the most vulnerable populations.

1.4 Climate Change and Air Pollution

Climate change and air pollution are closely linked, with climate change exacerbating existing air quality issues through a variety of processes. Rising global temperatures, altered weather patterns, and increased frequency of extreme weather events all contribute to the intensification of air pollution. These changes not only increase the concentration of harmful pollutants but also create conditions that extend their geographical impact and duration. Understanding these interactions is critical for addressing the combined challenges of global warming and deteriorating air quality, which pose significant risks to public health, including brain health [29] (Fig. 1.1).

Climate change exacerbates air pollution by increasing ground-level O_3 formation and PM concentrations [30]. With rising global temperatures, these chemical reactions intensify, especially in urban areas, worsening air quality. Heatwaves further amplify O_3 formation by creating stagnant air conditions, particularly during summer [1, 3, 30]. Exposure to elevated O_3 levels is linked to respiratory problems, which can in turn, effect brain health. Climate change also increases the concentration of fine PM2.5, which can penetrate the lungs and bloodstream, causing severe health impacts [8, 31]. The growing intensity and frequency of wildfires, a major source of PM2.5, contribute significantly to air pollution [31]. These fires release fine particles, soot (elemental and organic carbon), and toxic compounds that can travel substantial distances from their source, affecting air quality in distant regions [32]. Prolonged droughts and extreme heat also fuel these fires, prolonging wildfire seasons and further worsening air quality. Wildfire-generated PM2.5 is linked to an increase in hospital admissions for respiratory and cardiovascular issues, and has been associated with brain health issues including dementia [8, 33].

Furthermore, changes in precipitation patterns due to climate change affect air pollution. Droughts reduce the natural “cleansing” effect of rainfall, which usually helps wash pollutants from the atmosphere. Alternatively, excessive rainfall can mobilize ground pollutants, increasing the release of volatile compounds into the air [34]. The increased use of air conditioning in response to higher temperatures also raises energy demand, leading to more pollutant emissions from fossil fuel-based power plants. Agricultural practices, such as fertilizer and pesticide application, contribute to air pollution, with climate change influencing the timing and intensity of these activities [35].

1.5 Links Between Air Pollution and Neurological Diseases

In recent years, a growing body of evidence has established a clear association between air pollution and the development and progression of neurological diseases [36]. While air pollution has been known for its detrimental effects on respiratory and cardiovascular health, research now shows that PM2.5, O_3, NO_x, and other pollutants can also cross the blood-brain barrier, leading to neuroinflammation, oxidative stress, and neurodegenerative changes [8, 37, 38]. These processes have been implicated in the pathogenesis of several neurological disorders, including Alzheimer’s disease, Parkinson’s disease, stroke, and cognitive decline [39, 40] (Fig. 1.1).

1.5.1 Alzheimer’s Disease and Dementia

Alzheimer’s disease, the most common cause of dementia, is characterized by the accumulation of amyloid-beta plaques and tau tangles in the brain, leading to the progressive loss of memory, cognitive function, and independence. Emerging research suggests that long-term exposure to air pollution, particularly PM2.5 and NOx, may increase the risk of Alzheimer’s disease and accelerate cognitive decline in aging populations [41]. Several mechanisms have been proposed to explain the link between air pollution and Alzheimer’s disease. PM2.5 has been shown to enter the brain through olfactory pathways or by crossing the blood-brain barrier, where it can trigger neuroinflammation and oxidative stress [22]. This chronic inflammation can contribute to the accumulation of amyloid-beta plaques and the phosphorylation of tau [22]. Studies in both humans and animal models have demonstrated that individuals exposed to higher levels of PM2.5 have increased brain amyloid deposition and are at a higher risk for developing Alzheimer’s and other forms of dementia [42, 43]. Furthermore, epidemiological studies have shown that individuals living in areas with high levels of traffic-related air pollution, such as near highways or in densely populated urban centers, are more likely to develop dementia compared to those in cleaner environments [44]. These findings suggest a significant environmental component in the development of Alzheimer's disease, making air quality an important factor in the prevention of neurodegenerative diseases [45].

1.5.2 *Parkinson's Disease*

Parkinson's disease is a progressive neurodegenerative disorder primarily affecting movement, caused by the degeneration of dopaminergic neurons in the substantia nigra region of the brain. While genetic factors play a key role in the development of Parkinson's disease, environmental exposures, including air pollution, have been increasingly recognized as important contributors to the disease. Exposure to airborne pollutants, especially PM2.5 and traffic-related pollutants like NOx and CO, has been linked to a higher risk of developing Parkinson's disease [46]. These pollutants can induce oxidative stress and inflammation in the brain, which can contribute to the degeneration of dopamine-producing neurons [47]. Studies have demonstrated that long-term exposure to high levels of air pollution is associated with an increased incidence of Parkinson's disease, particularly in older adults. In addition, some pollutants, such as metals and persistent organic pollutants that are often found in PM2.5, have been identified as neurotoxic agents that can exacerbate neurodegeneration [48]. The mechanistic link between air pollution and Parkinson's disease is thought to involve inflammation of the central nervous system (CNS) as well as mitochondrial dysfunction, both of which are critical in the progression of the disease [49].

1.5.3 *Stroke and Cerebrovascular Diseases*

Stroke, a leading cause of disability and death worldwide, has also been strongly linked to air pollution. Both ischemic stroke (caused by a blockage in blood flow to the brain) and hemorrhagic stroke (caused by bleeding in the brain) have been associated with exposure to elevated levels of PM2.5, NOx, O_3, and sulfur dioxide (SO_2) [50]. These pollutants contribute to cardiovascular diseases, which are major risk factors for stroke, but they also directly affect the brain and cerebrovascular system [51]. Air pollution has been shown to increase the risk of stroke through several mechanisms, including systemic inflammation, oxidative stress, and increased blood coagulation [52]. PM2.5, in particular, is known to induce endothelial dysfunction, leading to the narrowing of blood vessels and an increased likelihood of clot formation. Additionally, chronic exposure to air pollution is associated with elevated blood pressure and increased arterial stiffness, both of which contribute to a heightened risk of stroke. Numerous studies have found a clear link between short-term increases in air pollution and an elevation in hospital admissions and mortality from stroke, highlighting the immediate impact of air quality on cerebrovascular health [53].

1.5.4 Cognitive Decline and Neurodevelopmental Disorders

In addition to adult neurodegenerative diseases, air pollution has been increasingly linked to cognitive decline. Long-term exposure to air pollutants such as PM2.5 and O_3 has been associated with a decline in cognitive function, memory loss, and impaired executive function [54]. The neuroinflammatory and oxidative stress mechanisms triggered by air pollution are thought to accelerate age-related cognitive decline and may increase the risk of developing mild cognitive impairment (MCI), a precursor to dementia [55]. Moreover, air pollution has been shown to affect brain development in children, leading to concerns about its role in neurodevelopmental disorders [56]. Prenatal and early-life exposure to high levels of air pollution has been linked to reduced IQ, attention deficits, and an increased risk of neurodevelopmental conditions such as ASD and ADHD [57]. PM and traffic-related pollutants are particularly harmful to the developing brain, as they can disrupt critical developmental processes and lead to long-lasting changes in brain structure and function [58].

1.5.5 Structural Brain Changes

Long-term exposure to air pollution can cause structural changes in the brain, including cortical thinning, reduced white matter integrity, and brain atrophy [59]. Magnetic resonance imaging (MRI) studies have shown that individuals exposed to higher levels of PM2.5 and NOx have smaller brain volumes and reduced connectivity in regions involved in higher-order cognitive functions [60]. These structural changes are thought to underlie the cognitive deficits observed in individuals exposed to chronic air pollution, contributing to memory loss, attention deficits, and executive dysfunction [60].

1.5.6 Mechanisms of Air Pollution-Induced Neurological Damage

The mechanisms by which air pollution affects the brain and increases the risk of neurological diseases are complex and still being elucidated. However, several key pathways have been identified [22, 61, 62]:

Neuroinflammation: Air pollutants such as PM2.5, O_3, and NOx can induce systemic inflammation, which leads to neuroinflammation in the brain. This inflammation can result in the activation of microglia (the brain's resident immune cells), which release pro-inflammatory cytokines and chemokines, contributing to neurodegenerative processes.

Oxidative Stress: Airborne pollutants are potent generators of reactive oxygen species (ROS), which can cause oxidative damage to neurons and other brain cells. Oxidative stress has been implicated in the pathogenesis of many neurological diseases, including Alzheimer's disease, Parkinson's disease, and stroke.

Blood-Brain Barrier Disruption: Fine particulate matter, particularly PM2.5, can cross the blood-brain barrier and accumulate in brain tissue. This can lead to direct damage to neurons and glial cells and exacerbate inflammation and oxidative stress within the brain.

Vascular Dysfunction: Air pollution contributes to systemic vascular inflammation and endothelial dysfunction, increasing the risk of cerebrovascular diseases such as stroke. These vascular changes can also reduce blood flow to the brain, impairing cognitive function and contributing to the progression of neurodegenerative diseases.

1.6 Acute Impacts of Air Pollution on Brain Function

Air pollution also induces acute effects on brain function, with its harmful effects spanning across various neurological areas. Exposure to pollutants such as PM2.5, O_3, and NO_x can lead to immediate, short-term neurological changes (as well as chronic, long-lasting damage) that increase the risk of cognitive decline and neurodegenerative diseases [63]. The brain's vulnerability to these pollutants is particularly concerning given its high energy demands and sensitivity to oxidative stress and inflammation, which are key pathological processes triggered by polluted air [64].

Cognitive Impairment: Even brief exposure to air pollution, particularly PM2.5 and O_3, has been shown to impair cognitive function. Studies indicate that people exposed to elevated pollution levels for a short period may experience slowed reaction times, decreased attention, and impaired memory. These effects are believed to be the result of rapid neuroinflammatory responses and oxidative stress triggered by pollutants entering the brain [65].

Psychological Health: Acute exposure to high levels of air pollution has also been linked to increased symptoms of anxiety, depression, and stress. Airborne pollutants can interfere with neurotransmitters, such as serotonin and dopamine pathways, which are essential for regulating mood and emotional stability. Studies have shown that individuals living in urban areas with high pollution levels report higher levels of psychological distress and are more likely to suffer from acute mood disturbances during pollution peaks [66].

Sleep Disturbances: Short-term exposure to air pollution can disrupt sleep quality, with individuals experiencing more difficulty falling asleep and maintaining restful sleep during periods of poor air quality. This is likely due to the inflammatory and neurotoxic effects of pollutants on the CNS, which can disturb normal sleep patterns and circadian rhythms [67].

Increased Risk of Stroke: As above, acute spikes in air pollution, particularly from PM2.5 and NO_x, have been associated with a rapid increase in the risk of ischemic stroke. The inflammatory and oxidative stress responses triggered by air pollution can contribute to endothelial dysfunction, platelet activation, and increased blood coagulation, leading to the formation of blood clots and restricted blood flow to the brain [68].

1.7 Heat, Air Quality, and Neurological Stress

The combined impacts of heat and poor air quality present a growing concern for brain health, especially in the context of climate change. Both factors, independently and synergistically, contribute to significant physiological stress, neuroinflammation, and cognitive impairments [8]. As heatwaves become more frequent and severe, and air quality deteriorates due to increased emissions and wildfires, understanding their combined effects on neurological health is essential for developing effective public health strategies [4].

1.7.1 Synergistic Effects of Heat and Poor Air Quality on Brain Health

The simultaneous exposure to high ambient temperatures and poor air quality creates a "double-hit" scenario for the brain, where the combined effects are more damaging than either factor alone [69]. Heat and air pollution interact to exacerbate systemic inflammation, oxidative stress, and neurotoxic processes that can impair cognitive function and increase the risk of neurological disorders [8, 69].

1.7.2 Heat-Induced Vulnerability to Air Pollution

When exposed to high temperatures, the body undergoes several physiological changes to maintain homeostasis, such as increased heart rate, elevated body temperature, and vasodilation. These changes can weaken the body's defenses against air pollutants and increase the absorption of toxic compounds, including PM2.5 and PM10, into the bloodstream and brain.

1.7.3 *Increased Inhalation of Pollutants*

Heat-induced hyperventilation can cause individuals to breathe more rapidly and deeply, increasing their intake of air pollutants. This heightened exposure to pollutants like PM2.5, O_3, and NOx during heat waves amplifies the harmful effects of air quality on brain health. Prolonged exposure to polluted air in hot conditions can accelerate the transport of pollutants into the lungs and circulation, ultimately reaching the brain, where they induce neuroinflammation and oxidative stress [70].

1.7.4 *Compromised Blood-Brain Barrier (BBB)*

The BBB, which protects the brain from harmful substances in the bloodstream, is more likely to become compromised in high-heat environments. Studies have shown that both heat stress and PM can independently weaken the integrity of the BBB. When these factors co-occur, they may further disrupt the BBB, allowing neurotoxic particles, such as PM2.5, to enter the brain more readily, contributing to long-term cognitive damage and neurodegenerative diseases [71].

1.8 Conclusion

A growing body of evidence supports the significant impact of climate change on air pollution and heat and their subsequent effects on neurological health. Chronic exposure to air pollutants like PM2.5, O_3, and NO*x* is linked to increased neuroinflammation, oxidative stress, and structural changes in the brain, all of which contribute to conditions such as developmental brain impairment, Alzheimer's disease, Parkinson's disease, and stroke. Acute exposure, especially during high pollution events, can lead to immediate cognitive impairments and psychological disturbances, emphasizing the urgent need for public health interventions (Fig. 1.1).

As climate change continues to escalate these environmental stressors, understanding their combined effects is crucial for formulating effective strategies to protect brain health. This complex challenge requires a concerted effort from researchers, policymakers, and public health officials to mitigate the risks associated with air pollution and climate change, ultimately protecting neurological health for current and future generations.

References

1. Fiore, A. M., Naik, V., & Leibensperger, E. M. (2015). Air quality and climate connections. *Journal of the Air & Waste Management Association (1995), 65*, 645.
2. Orru, H., Ebi, K. L., & Forsberg, B. (2017). The interplay of climate change and air pollution on health. *Current Environmental Health Reports, 4*, 504.
3. Jacob, D. J., & Winner, D. A. (2009). Effect of climate change on air quality. *Atmospheric Environment, 43*, 51–63.
4. Louis, S., Carlson, A. K., Suresh, A., Rim, J., Mays, M., Ontaneda, D., & Dhawan, A. (2023). Impacts of climate change and air pollution on neurologic health, disease, and practice: A scoping review. *Neurology, 100*, 474.
5. Tang, X., Wilson, S. R., Solomon, K. R., Shao, M., & Madronich, S. (2011). Changes in air quality and tropospheric composition due to depletion of stratospheric ozone and interactions with climate. *Photochemical & Photobiological Sciences, 10*, 280.
6. Orellano, P., Reynoso, J., Quaranta, N., Bardach, A., & Ciapponi, A. (2020). Short-term exposure to particulate matter (PM_{10} and $PM_{2.5}$), nitrogen dioxide (NO_2), and ozone (O_3) and all-cause and cause-specific mortality: Systematic review and meta-analysis. *Environment International, 142*, 105876.
7. Xu, R., Yu, P., Abramson, M. J., Johnston, F. H., Samet, J. M., Bell, M. L., Haines, A., Ebi, K. L., Li, S., & Guo, Y. (2020). Wildfires, global climate change, and human health. *The New England Journal of Medicine, 383*, 2173.
8. White, A. R. (2024). The firestorm within: A narrative review of extreme heat and wildfire smoke effects on brain health. *Science of the Total Environment, 922*, 171239.
9. Afifa, K., Arshad, N., Hussain, M. H., Ashraf, M. Z., & Saleem. (2024). Air pollution and climate change as grand challenges to sustainability. *Science of the Total Environment, 928*, 172370.
10. Lelieveld, J., Evans, J. S., Fnais, M., Giannadaki, D., & Pozzer, A. (2015). The contribution of outdoor air pollution sources to premature mortality on a global scale. *Nature, 525*, 367.
11. Turner, M. C., Jerrett, M., Pope, C. A., Krewski, D., Gapstur, S. M., Diver, W. R., Beckerman, B. S., Marshall, J. D., Su, J., Crouse, D. L., & Burnett, R. T. (2016). Long-term ozone exposure and mortality in a large prospective study. *American Journal of Respiratory and Critical Care Medicine, 193*, 1134.
12. Middleton, N. (2018). Rangeland management and climate hazards in drylands: Dust storms, desertification and the overgrazing debate. *Natural Hazards, 92*, 57–70.
13. Kelly, F. J., & Fussell, J. C. (2020). Global nature of airborne particle toxicity and health effects: A focus on megacities, wildfires, dust storms and residential biomass burning. *Toxicology Research (Camb), 9*, 331.
14. Smith, M. I., Ke, Y., Geyman, E. C., Reahl, J. N., et al. (2024). Mercury stocks in discontinuous permafrost and their mobilization by river migration in the Yukon River Basin. *Environmental Research Letters, 19*(8), 084041.
15. Kenny, G. P., Yardley, J., Brown, C., Sigal, R. J., & Jay, O. (2010). Heat stress in older individuals and patients with common chronic diseases. *CMAJ, 182*, 1053.
16. Piscitelli, P., & Miani, A. (2024). Climate change and infectious diseases: Navigating the intersection through innovation and interdisciplinary approaches. *International Journal of Environmental Research and Public Health, 21*, 314.
17. Mirzabaev, A., Kerr, R. B., Hasegawa, T., & Pradhan, P. (2023). Severe climate change risks to food security and nutrition. *Climate Risk Management, 39*, 100473.
18. Clayton, S. (2021). Climate change and mental health. *Current Environmental Health Reports, 8*, 1.
19. van Nieuwenhuizen, A., Hudson, K., Chen, X., & Hwong, A. R. (2021). The effects of climate change on child and adolescent mental health: Clinical considerations. *Current Psychiatry Reports, 23*, 88.

20. Scattolin, M. A. A., Resegue, R. M., & Rosário, M. C. D. (2022). The impact of the environment on neurodevelopmental disorders in early childhood. *Jornal de Pediatria, 98*(Suppl 1), S66.
21. Manisalidis, I., Stavropoulou, E., Stavropoulos, A., & Bezirtzoglou, E. (2020). Environmental and health impacts of air pollution: A review. *Frontiers in Public Health, 8*, 14.
22. Fu, C., Kuang, D., Zhang, H., Ren, J., & Chen, J. (2022). Different components of air pollutants and neurological disorders. *Frontiers in Public Health, 10*, 959921.
23. Li, Y., Xie, T., Cardoso Melo, R. D., de Vries, M., Lakerveld, J., Zijlema, W., & Hartman, C. A. (2023). Longitudinal effects of environmental noise and air pollution exposure on autism spectrum disorder and attention-deficit/hyperactivity disorder during adolescence and early adulthood: The TRAILS study. *Environmental Research, 227*, 115704.
24. Bongioanni, P., Del Carratore, R., Corbianco, S., Diana, A., Cavallini, G., Masciandaro, S. M., Dini, M., & Buizza, R. (2021). Climate change and neurodegenerative diseases. *Environmental Research, 201*, 111511.
25. Junck, L., Saad, A., & Andrews, B. (2024). Neurology and climate change. *Lancet Neurology, 23*, 552.
26. Iqubal, A., Ahmed, M., Ahmad, S., Sahoo, C. R., Iqubal, M. K., & Haque, S. E. (2020). Environmental neurotoxic pollutants: Review. *Environmental Science and Pollution Research International, 27*, 41175.
27. Kusters, M. S. W., López-Vicente, M., Muetzel, R. L., et al. (2024). Residential ambient air pollution exposure and the development of white matter microstructure throughout adolescence. *Environmental Research, 262*(Pt 2), 119828.
28. Ribas-Fitó, N., Sala, M., Kogevinas, M., & Sunyer, J. (2001). Polychlorinated biphenyls (PCBs) and neurological development in children: A systematic review. *Journal of Epidemiology and Community Health, 55*, 537.
29. Ebi, K. L., Vanos, J., Baldwin, J. W., Bell, J. E., Hondula, D. M., Errett, N. A., Hayes, K., Reid, C. E., Saha, S., Spector, J., & Berry, P. (2021). Extreme weather and climate change: Population health and health system implications. *Annual Review of Public Health, 42*, 293.
30. Vilcins, D., Christofferson, R. C., Yoon, J. H., Nazli, S. N., Sly, P. D., Cormier, S. A., & Shen, G. (2024). Updates in air pollution: Current research and future challenges. *Annals of Global Health, 90*, 9.
31. Liu, J. C., Mickley, L. J., Sulprizio, M. P., Dominici, F., Yue, X., Ebisu, K., Anderson, G. B., Khan, R. F. A., Bravo, M. A., & Bell, M. L. (2016). Particulate air pollution from wildfires in the Western US under climate change. *Climatic Change, 138*, 655.
32. Woo, S. H. L., Liu, J. C., Yue, X., Mickley, L. J., & Bell, M. L. (2020). Air pollution from wildfires and human health vulnerability in Alaskan communities under climate change. *Environmental Research Letters, 15*, 094019.
33. Krawisz, B. (2020). Health effects of climate destabilization: Understanding the problem. *WMJ, 119*, 132.
34. Bolan, S., Padhye, L. P., Jasemizad, T., Govarthanan, M., Karmegam, N., Wijesekara, H., Amarasiri, D., Hou, D., Zhou, P., Biswal, B. K., Balasubramanian, R., Wang, H., Siddique, K. H. M., Rinklebe, J., Kirkham, M. B., & Bolan, N. (2024). Impacts of climate change on the fate of contaminants through extreme weather events. *Science of the Total Environment, 909*, 168388.
35. Aneja, V. P., Schlesinger, W. H., & Erisman, J. W. (2009). Effects of agriculture upon the air quality and climate: Research, policy, and regulations. *Environmental Science & Technology, 43*, 4234.
36. Russ, T. C., Reis, S., & van Tongeren, M. (2019). Air pollution and brain health: Defining the research agenda. *Current Opinion in Psychiatry, 32*, 97.
37. Milton, L. A., & White, A. R. (2020). The potential impact of bushfire smoke on brain health. *Neurochemistry International, 139*, 104796.
38. Haghani, A., Morgan, T. E., Forman, H. J., & Finch, C. E. (2020). Air pollution neurotoxicity in the adult brain: Emerging concepts from experimental findings. *Journal of Alzheimer's Disease, 76*, 773.

39. Costa, L. G., Cole, T. B., Dao, K., Chang, Y. C., & Garrick, J. M. (2019). Developmental impact of air pollution on brain function. *Neurochemistry International, 131*, 104580.
40. Calderón-Garcidueñas, L., & Ayala, A. (2022). Air pollution, ultrafine particles, and your brain: Are combustion nanoparticle emissions and engineered nanoparticles causing preventable fatal neurodegenerative diseases and common neuropsychiatric outcomes. *Environmental Science & Technology, 56*, 6847.
41. Ma, H., Li, H., Zhang, Y., Zhou, Y., Liu, H., Xu, H., Zhu, L., Zhang, G., Wang, J., Li, Z., Hong, B., Zhou, W., Yang, P., & Liu, J. (2023). Microglia exhibit distinct heterogeneity rather than M1/M2 polarization within the early stage of acute ischemic stroke. *Aging and Disease, 14*(6), 2284–2302.
42. Fu, P., & Yung, K. K. L. (2020). Air pollution and Alzheimer's disease: A systematic review and meta-analysis. *Journal of Alzheimer's Disease, 77*, 701.
43. Kilian, J., & Kitazawa, M. (2018). The emerging risk of exposure to air pollution on cognitive decline and Alzheimer's disease—Evidence from epidemiological and animal studies. *Biomedical Journal, 41*, 141.
44. Costa, L. G., Cole, T. B., Coburn, J., Chang, Y. C., Dao, K., & Roqué, P. J. (2017). Neurotoxicity of traffic-related air pollution. *Neurotoxicology, 59*, 133.
45. Tham, R., & Schikowski, T. (2021). The role of traffic-related air pollution on neurodegenerative diseases in older people: An epidemiological perspective. *Journal of Alzheimer's Disease, 79*, 949.
46. Murata, H., Barnhill, L. M., & Bronstein, J. M. (2022). Air pollution and the risk of Parkinson's disease: A review. *Movement Disorders, 37*, 894.
47. Palacios, N. (2017). Air pollution and Parkinson's disease—Evidence and future directions. *Reviews on Environmental Health, 32*, 303.
48. Goldman, S. M. (2014). Environmental toxins and Parkinson's disease. *Annual Review of Pharmacology and Toxicology, 54*, 141.
49. Block, M. L., & Calderón-Garcidueñas, L. (2009). Air pollution: Mechanisms of neuroinflammation and CNS disease. *Trends in Neurosciences, 32*, 506.
50. Verhoeven, J. I., Allach, Y., Vaartjes, I. C. H., Klijn, C. J. M., & de Leeuw, F. E. (2021). Ambient air pollution and the risk of ischaemic and haemorrhagic stroke. *Lancet Planetary Health, 5*, e542.
51. Lee, K. K., Miller, M. R., & Shah, A. S. V. (2018). Air pollution and stroke. *Journal of Stroke, 20*, 2.
52. Kulick, E. R., Kaufman, J. D., & Sack, C. (2023). Ambient air pollution and stroke: An updated review. *Stroke, 54*, 882.
53. Hahad, O., Lelieveld, J., Birklein, F., Lieb, K., Daiber, A., & Münzel, T. (2020). Ambient air pollution increases the risk of cerebrovascular and neuropsychiatric disorders through induction of inflammation and oxidative stress. *International Journal of Molecular Sciences, 21*, 4306.
54. Paul, K. C., Haan, M., Mayeda, E. R., & Ritz, B. R. (2019). Ambient air pollution, noise, and late-life cognitive decline and dementia risk. *Annual Review of Public Health, 40*, 203.
55. Schikowski, T., & Altuğ, H. (2020). The role of air pollution in cognitive impairment and decline. *Neurochemistry International, 136*, 104708.
56. Costa, L. G., Cole, T. B., Dao, K., Chang, Y. C., Coburn, J., & Garrick, J. M. (2020). Effects of air pollution on the nervous system and its possible role in neurodevelopmental and neurodegenerative disorders. *Pharmacology & Therapeutics, 210*, 107523.
57. Lyall, K., Croen, L., Daniels, J., Fallin, M. D., Ladd-Acosta, C., Lee, B. K., Park, B. Y., Snyder, N. W., Schendel, D., Volk, H., Windham, G. C., & Newschaffer, C. (2017). The changing epidemiology of autism spectrum disorders. *Annual Review of Public Health, 38*, 81.
58. Allen, J. L., Oberdorster, G., Morris-Schaffer, K., Wong, C., Klocke, C., Sobolewski, M., Conrad, K., Mayer-Proschel, M., & Cory-Slechta, D. A. (2017). Developmental neurotoxicity of inhaled ambient ultrafine particle air pollution: Parallels with neuropathological and behavioral features of autism and other neurodevelopmental disorders. *Neurotoxicology, 59*, 140.

59. Bos, B., Barratt, B., Batalle, D., Gale-Grant, O., Hughes, E. J., Beevers, S., Cordero-Grande, L., Price, A. N., Hutter, J., Hajnal, J. V., Kelly, F. J., David Edwards, A., & Counsell, S. J. (2023). Prenatal exposure to air pollution is associated with structural changes in the neonatal brain. *Environment International, 174*, 107921.
60. Yuan, A., Halabicky, O., & Liu, J. (2024). Association between air pollution exposure and brain cortical thickness throughout the lifespan: A systematic review. *Neuroscience, 559*, 209.
61. Kim, H., Kim, W. H., Kim, Y. Y., & Park, H. Y. (2020). Air pollution and central nervous system disease: A review of the impact of fine particulate matter on neurological disorders. *Frontiers in Public Health, 8*, 575330.
62. Genc, S., Zadeoglulari, Z., Fuss, S. H., & Genc, K. (2012). The adverse effects of air pollution on the nervous system. *Journal of Toxicology, 2012*, 782462.
63. Milani, C., Farina, F., Botto, L., Massimino, L., Lonati, E., Donzelli, E., Ballarini, E., Crippa, L., Marmiroli, P., Bulbarelli, A., & Palestini, P. (2020). Systemic exposure to air pollution induces oxidative stress and inflammation in mouse brain, contributing to neurodegeneration onset. *International Journal of Molecular Sciences, 21*, 3699.
64. de Prado Bert, P., Mercader, E. M. H., Pujol, J., Sunyer, J., & Mortamais, M. (2018). The effects of air pollution on the brain: A review of studies interfacing environmental epidemiology and neuroimaging. *Current Environmental Health Reports, 5*, 351.
65. Gao, X., Coull, B., Lin, X., Vokonas, P., Spiro, A., Hou, L., Schwartz, J., & Baccarelli, A. A. (2021). Short-term air pollution, cognitive performance, and nonsteroidal anti-inflammatory drug use in the Veterans Affairs Normative Aging Study. *Nature Aging, 1*, 430.
66. Qiu, X., Danesh-Yazdi, M., Wei, Y., Di, Q., Just, A., Zanobetti, A., Weisskopf, M., Dominici, F., & Schwartz, J. (2022). Associations of short-term exposure to air pollution and increased ambient temperature with psychiatric hospital admissions in older adults in the USA: A case-crossover study. *Lancet Planetary Health, 6*, e331.
67. Ayta, S. (2024). The effect of air pollution and climate change on sleep. *Noro Psikiyatri Arsivi, 61*, 73.
68. Shah, A. S., Lee, K. K., McAllister, D. A., Hunter, A., Nair, H., Whiteley, W., Langrish, J. P., Newby, D. E., & Mills, N. L. (2015). Short term exposure to air pollution and stroke: Systematic review and meta-analysis. *BMJ, 350*, h1295.
69. Zhou, W., Wang, Q., Li, R., Zhang, Z., Wang, W., Zhou, F., & Ling, L. (2023). The effects of heatwave on cognitive impairment among older adults: Exploring the combined effects of air pollution and green space. *Science of the Total Environment, 904*, 166534.
70. Rai, M., Stafoggia, M., de'Donato F, Scortichini, M., Zafeiratou, S., Vazquez Fernandez, L., Zhang, S., Katsouyanni, K., Samoli, E., Rao, S., Lavigne, E., Guo, Y., Kan, H., Osorio, S., Kyselý, J., Urban, A., Orru, H., Maasikmets, M., Jaakkola, J. J. K., Ryti, N., Pascal, M., Hashizume, M., Fook Sheng Ng, C., Alahmad, B., Hurtado Diaz, M., De la Cruz Valencia, C., Nunes, B., Madureira, J., Scovronick, N., Garland, R. M., Kim, H., Lee, W., Tobias, A., Íñiguez, C., Forsberg, B., Åström, C., Maria Vicedo-Cabrera, A., Ragettli, M. S., Leon Guo, Y. L., Pan, S. C., Li, S., Gasparrini, A., Sera, F., Masselot, P., Schwartz, J., Zanobetti, A., Bell, M. L., Schneider, A., & Breitner, S. (2023). Heat-related cardiorespiratory mortality: Effect modification by air pollution across 482 cities from 24 countries. *Environment International, 174*, 107825.
71. Kiyatkin, E. A., & Sharma, H. S. (2009). Permeability of the blood-brain barrier depends on brain temperature. *Neuroscience, 161*, 926.

Chapter 2
Ambient Air Pollution: Which Components Are Most Important for the Brain?

Pasi Jalava, Henri Hakkarainen, and Topi Rönkkö

Keywords Air pollution · Ultrafine particles · Physicochemical composition · Brain health

2.1 Introduction

Air pollution is considered one of the main environmental threats that negatively impact human health. Depending on the estimates, the current burden of premature mortality is counted in several millions [59]. Exposure to air pollution may also compromise brain health; however, knowledge on this topic is still insufficient for future regulatory purposes. Ambient particulate matter (PM) is considered the most harmful component of air pollution, but gaseous pollutants may also pose a threat to human health, directly or via their contribution to atmospheric secondary aerosol formation. Indoor pollution is additionally important when considering the adverse effects caused by air pollutants. Although the negative health effects of indoor pollutants are currently strongly concentrated in the developing areas of the globe, mostly due to solid and liquid fuel combustion indoors with poor ventilation [118], also the infiltration of outdoor pollution to indoors in developed societies should be considered when assessing the health impacts of air pollution (e.g. [105]).

The air pollution levels are very unevenly distributed globally, with the majority of people exposed to unsafe levels of air pollution living in low- and middle-income countries [89]. Over 90 per cent of the global population lives in areas where the guideline limits are surpassed, with the highest levels found in developing countries

P. Jalava (✉) · H. Hakkarainen
Inhalation toxicology laboratory, Department of Environmental and Biological Sciences, University of Eastern Finland, Kuopio, Finland
e-mail: pasi.jalava@uef.fi

T. Rönkkö
Aerosol Physics Laboratory, Physics Unit, Faculty of Engineering and Natural Sciences, Tampere University, Tampere, Finland

K. M. Kanninen, A. R. White (eds.), *Air Pollution and the Brain*,
https://doi.org/10.1007/978-3-031-99302-2_2

where combustion-based technology is usually old and residential cooking and heating are grounded on indoor combustion events [96, 104, 118]. Moreover, in Europe, there are still relatively large differences in the emissions and air quality both in the north-south and east-west axes [28]. Air pollution levels depend on multiple anthropogenic and natural sources such as industry, traffic, and wildfires (Fig. 2.1). The chemical and physical composition of air pollution also depends on the area within the globe; in addition to the sources, environmental factors such as meteorological conditions, such as humidity and temperature affect the emissions [117]. Especially secondary aerosols are affected by these, and epidemiological studies have shown that the health effects of air pollutants depend on the season [63]. Different countries and regions have implemented differing air pollution controls and limit values. For example, several air pollution regulations have focused on mitigating traffic exhaust emissions [90]. World Health Organization (WHO) guidelines for ambient air pollution are currently met only in a few places in the world. With the advent of the new, more strict guideline values, the target levels in many locations are even further away. Overall, the varying air pollution levels affect air quality and human health differently across regions.

Traditionally, particulate air pollution is connected mostly to respiratory and cardiovascular health. However, recently, air pollutants have been linked to various other adverse effects, including those in the brain. In this chapter, we aim to shed light on different air pollution sources as well as pollutants and their properties that

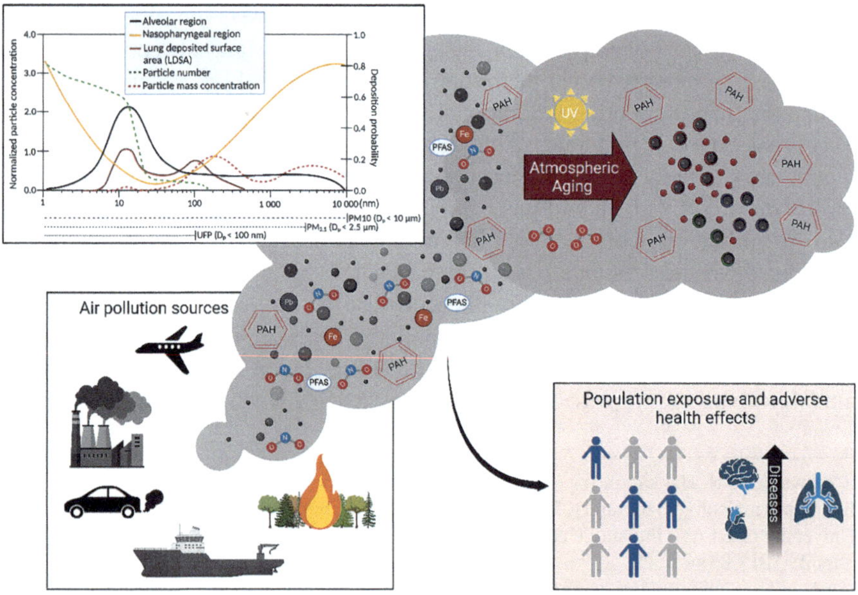

Fig. 2.1 Different anthropogenic sources, such as traffic, industry, energy production, and wildfires. Atmospheric aging processes result in the production of secondary aerosols and the absorption of various compounds onto the particle surfaces. Deposition curves in parts of the respiratory tract for the differently sized PM mass, number, and lung-deposited surface area

may have harmful implications for human health, especially for the brain. Recent studies suggest emerging evidence linking air pollution exposure to an increased risk of dementia, particularly in correlation with traffic emissions [81]. The components of air pollution that have been most associated with brain effects are the smallest particulate matter fractions, fine ($PM_{2.5}$) and ultrafine particles (UFP) [40].

2.2 Size Ranges Relevant to Adverse Brain Health Affects

Particulate air pollution is typically monitored in the atmosphere by measuring coarse particles (PM_{10}) with particle size <10 μm and fine particles ($PM_{2.5}$) with particle size <2.5 μm. The mass concentrations of these particles are required to be measured in urban areas to protect human health and to obtain input to, e.g. air quality indexes. UFPs with particle sizes of 100 nm or less have not been included in air quality legislation, even though their potential for adverse health effects has been recognized. In the most recent WHO air quality guidelines, however, there is a strong recommendation to measure both UFP and black carbon (BC) in the air to get better scientific evidence on their presence in the air as well as their adverse health effects [118]. Very recently, the European Union accepted the new air quality directive, where for the first time, there is a requirement to measure UFPs in the atmosphere as particle number concentration (PNC) and size distribution [26].

Ambient aerosol particles are externally and internally mixed, meaning that the particle populations consist of differently sized particles and that the individual particles can contain several elements and compounds. The size distribution of ambient particles typically has several modes that originate from different particle formation processes, e.g. so-called nucleation mode particles are formed by atmospheric gas-to-particle processes, but the soot mode particles form in combustion processes. After their emission to the atmosphere or formation in the atmosphere, particles are exposed to atmospheric processes that can significantly alter their properties, and size, affecting their concentrations [93]. Both the initial formation mechanisms and the atmospheric processes affect the size, composition, and morphology of the particles, and those parameters can have effects on the harmfulness of the particles to human health. On the other hand, the complexity of aerosol sources and atmospheric processes affecting the aerosol properties of aerosol populations results in large spatial and temporal variations in ambient aerosols, especially in urban areas, which significantly increases the challenges in the evaluation of health impacts of the aerosols.

Biologically, particle size is one of the greatest determinants of where the particles end up in the body. Generally, larger particles are deposited in the upper airway, and the smaller the particles are, the deeper they can penetrate the respiratory tract. In addition, the very smallest UFPs deposit effectively in the head airways due to diffusion. It is also known that UFPs can be transported to the blood circulation from the lungs, thus potentially affecting every organ in the body [78]. Measurements of atmospheric concentrations of UFPs are commonly expressed as particle number

(PN) per unit volume [93]. Moreover, metrics such as lung-deposited surface area (LDSA) and oxidative potential (OP) can be gradually used to assess the probable deposition of the particulates in the respiratory tract and the potential of the particulates to induce oxidate stress in biological systems, respectively [60, 74].

2.2.1 Ultrafine Particles

As mentioned above, one important factor in health impact evaluations is the deposition efficiency of the particles in the human respiratory tract and in its different parts. From the distinct particle size fractions, the UFPs have been associated mostly with adverse health effects on the brain. In general, most of the smallest UFPs of inhaled air, sub-10 nm particles, are deposited onto the surfaces of the upper human airways, which may enable a direct route of these particles to the brain via the olfactory route. Meanwhile, particles in the 10–100 nm size range are deposited significantly in the alveolar region of human lungs, where they are potentially transported through an epithelial barrier to blood circulation [76]. Some studies have even suggested that UFP-sized particulates can penetrate the blood-brain barrier, resulting in an additional route of deposition from the circulation to the brain [40]. Among other organs, the brain can also be affected by systemic inflammation caused by exposure to UFPs [73]. Therefore, due to their properties and deposition efficiency, ultrafine particles are the most probable driver of adverse effects in the brain.

In urban areas, UFPs typically dominate the total particle number concentrations of ambient air. However, the UFP has, on average, a minor contribution to the mass concentration of particles of ambient air. Instead, the mass concentration of particles is governed by particles larger than 100 nm, often even by coarse particles larger than 2.5 μm. However, it should be kept in mind that the particle concentrations and particle size distributions have large variations, and both are affected by regional pollution levels, local emission sources, geographical factors, and atmospheric conditions. The particle size range from 10 to 100 nm can contain different types of particles, which is seen in their size distribution as different modes; the size range can have two or even three modes, (nucleation mode [<10 nm], Aitken and [10–50 nm], and soot mode [>50 nm]) all have different composition and mechanism of formation. For example, nucleation mode and Aitken mode particles often contain semivolatile compounds that can originate from human activities. Also, the so-called accumulation mode can contribute to this particle size range, although especially its contribution to particle mass can come mostly from particle sizes larger than 100 nm. It should be noted that the naming of the modes is not straightforward, and local emission sources and atmospheric conditions can affect the existence of each mode significantly. For example, soot or black carbon particles can mostly be found from this particle size range, i.e. from 10 to 100 nm, and together with associated semivolatile organic and inorganic compounds, their harmful contribution to human health has been indicated by toxicological and epidemiological studies [33, 109]. Moreover, concerning brain health, black carbon particles have been found in the human brain in regions that are important for memory [114], and they have been observed to cross the placental barrier [9] and have even been found in the fetal brain. However, there is currently little information on the role of semi-volatile components in brain health.

Particles in the sub-10 nm size range of ambient air can be in urban atmospheres in high number concentrations [43, 94]. Their concentrations or their emissions to the atmosphere are not directly regulated by any air quality or emission standards. Although these small particles may possess non-volatile compounds that can be carbonaceous or metallic, which are emitted to the atmosphere as primary particle emissions e.g., from traffic, the atmospheric formation from semivolatile compounds seems to be their dominant formation process. The atmospheric formation and growth processes of the particles are complex and linked with environmental conditions and the availability of atmospheric precursor gases; for instance, solar irradiation affects the formation of atmospheric radicals and the atmospheric concentrations of gaseous sulphuric acid molecules, and the concentration of sulphuric acid is seen important in new particle formation and growth at least in rural and urban background areas [106]. However, gaseous sulphuric acid is also emitted by combustion processes of sulphur-containing fossil fuels, which can then contribute to direct nanoparticle emissions from internal combustion engines [4, 79]. It should be noted that the atmospheric formation of sub-10 nm particles can be contributed to by several different compounds, and it is heavily affected by atmospheric conditions. Also, the amount and composition of sub-10 nm particles directly emitted to the atmosphere from various sources are largely unknown. This complicates the evaluation of the potential health impacts they can have.

Recent research indicates that traffic is a dominant source of sub-10 nm particles in urban environments [43, 94]. This is seen as high sub-10 nm particle concentrations near traffic routes, which are simultaneously the environments where people are commuting and living. This, together with the high deposition efficiency of the particles to the human respiratory tract (total deposition efficiency >90%, high deposition efficiency to head airways, tracheobronchial part, and alveolar part, in this order) makes them potentially very important contributors to brain diseases. Substantial head airway deposition also means the exposure of the olfactory bulb to these tiny particles [65]. Indeed, the smallest nanoparticles accumulate most efficiently in the human olfactory region, and the deposition in this region decreases towards 100 nm [29]. The translocation of ultrafine particles through the olfactory nerve is shown in animal models (e.g. [23, 45]). There is growing evidence on the ability of ultrafine particles to cross bodily membranes and reach even the brain [10, 14, 35, 39, 57, 93, 101]. This can partly be explained by the fact that the particles are small, also if compared to the dimensions of cellular structures of the olfactory bulb and epithelium of head airways in general, which can enable, in principle, efficient penetration of particles into the tissues.

In addition to particle size affecting the deposition of particles and translocation of particles in the human body, the composition and physical properties of UFPs also impact their potential on brain health. However, for the sub-10 nm sized particulates, due to their small particle size and negligible mass concentration compared to larger particles, information regarding the composition and properties of sub-10 nm particles is highly limited, especially since they are mixed within ambient aerosols. It has been observed that in urban traffic–influenced areas, a significant fraction of these particles contains a non-volatile fraction [64]. Based on the laboratory studies for individual vehicles or vehicle technologies, these tiny non-volatile particles, also called core particles of nucleation mode particles, can contain elemental carbon and metallic compounds formed in the particle phase during the combustion process [3, 21, 92, 103] and compounds from vehicle brakes [77]. On the other hand, an even more significant fraction of these particles can consist of semivolatile compounds [64], e.g., sulfuric acid [4], and hydrocarbons [82]. The overall composition of the UFPs affects the properties and health effects. For example, water solubility can influence the fate of the particles in the human body after they have been deposited into the airway system; the compounds with high water solubility are likely dissolved in the liquid film on the epithelium surface, and the effects of compounds are based on the molecular level properties of compounds. However, if the water solubility is low, they can remain as solid particles also after the deposition. In that case, their translocation in the body and the impacts can be assumed to differ from rapidly dissolved particles.

The size range from 10 to 100 nm can be efficiently affected by emission regulations; e.g. European emission standards have limited the number of emissions of non-volatile particles larger than 23 nm since the Euro 5/V standard came into force, and the aviation emission regulations cover non-volatile particles larger than 10 nm. In European cities, these standards have affected particle number concentration and black carbon concentrations of urban air, both in roadside environments and in urban backgrounds [20, 67]. In addition to limitations on the number of solid particle emissions, there are also mass-based emission limits for particles and restrictions on gaseous emissions that impact particles in the smaller size range. Emission limits for cars, heavy-duty vehicles, and non-road mobile machinery not only include particulate mass emission limits but also gaseous emission limits, both of which affect particles measuring between 10 and 100 nanometres. Despite this, the UFP concentrations have not decreased similarly compared to the larger particle size fractions, and there is still a lack of measurement data from UFPs [87]. Overall, due to their distinct properties, such as the high number-to-mass ratio, high surface area, chemical composition, and areas of deposition in the airway, ultrafine particles can be considered especially potent to cause harmful effects in the brain.

2.3 The Most Important Stationary Combustion Sources of Air Pollution

2.3.1 Large-Scale Combustion Units

The anthropogenic origins of air pollution include several different sources, mostly related to combustion on distinct scales. Overall, combustion is still a primary energy source in large regions worldwide, although other energy sources are becoming more common. It includes the combustion of solid, liquid, and gaseous fuels on scales ranging from large power plants to domestic combustion. From large-scale stationary sources, coal and oil combustion are the major contributors to air quality and harmful emissions [49]. However, in large power plants, the stack height is much higher than e.g. domestic combustion. Therefore, the impact on local air quality may be larger from small-scale combustion, and larger plants affect the background air quality in larger areas. Coal and oil combustion are used in electricity generation, industrial processes, and heating. Coal combustion is especially detrimental to air quality and health since the emissions contain high concentrations of harmful particulate matter, gaseous compounds, and PAHs [71]. On the other hand, the emissions of coal-fired power plants are typically strictly regulated in industrialized countries, and e.g. SO_2 and primary particle emissions have been diminished efficiently (e.g. [75, 99]). In the case of oil combustion, fuel quality plays a significant role in emissions. The different distillate fractions also have largely differing fuel compositions, which reflects the emissions. However, it should be kept in mind that the effects of fuel quality on the emissions can be complex; e.g. in the study of Happonen et al. [36] on the emissions from an oil-fired heating plant, a mixture of light fuel oil (LFO) and heavy fuel oil (HFO) fuels resulted in considerably higher black carbon content in particulate matter when compared to the combustion of HFO only. The sulphur content of the fuel is one of the major contributors to air quality in large-scale combustion of fossil oil and fossil solid fuel combustion, affecting also the atmospheric secondary aerosol formation (e.g. [5]), but the emissions contain in addition other harmful components, such as metal-containing ash [99]. However, in large power plants, those can be effectively filtered from the emissions of both oil and coal combustion. This is, unfortunately, not the case everywhere, and especially in developing regions, the purification of emissions is often not at the same level as in developed regions. Coal and oil combustion are also used in smaller industries or even in households (e.g. [95]), where there may not be any after-treatment of the emissions before releasing them into the atmosphere. The small-scale combustion can lead to local air pollution episodes, especially in winter, resulting in hazardous concentrations of the emissions [125].

2.3.2 *Small-Scale Combustion*

Another stationary source of air pollution is the activities of small-scale combustion, such as residential heating using fossil and renewable fuels. This includes the combustion of fuels such as biomass, which have been estimated to be an important source of air pollution in Europe alone [18, 46]. Moreover, biomass combustion appliances and practices are usually outdated, resulting in significantly higher emissions than modern devices [115]. Biomass emissions can result in air pollution episodes, especially during the winter months [97]. When compared to traffic emissions, biomass produces emissions with higher concentrations of organic carbon molecules, resulting in distinct atmospheric aging processes, secondary aerosols, and toxicological responses [62]. Biomass emissions have been shown to have a multi-fold higher oxidative potential even after several days, indicating the possible adverse health effects of these emissions after atmospheric transportation processes, thus resulting in adverse health effects far from the source [121].

2.4 Importance of Traffic Modes on Air Pollution

2.4.1 Road Traffic

In road traffic, distillate fuels are still very widely used despite the electrification of the traffic. Internal combustion engines still dominate the current fleet in most parts of the world, and in heavy-duty traffic, this will probably prevail for a long time to come [47]. Emission control regulations have been effective in reducing the exhaust particulate mass and black carbon concentrations in the ambient air (e.g. [67]), but the question remains whether the UFPs with low mass but high number concentrations have been proportionally decreased [20].

Living near major roads has been associated with structural effects on the brain and various neurological diseases [16, 120]. However, the exact reasons for the findings are not clear. Traffic produces overall a complex mixture of pollutants from combustion sources, and mechanical wear particles, with the addition of co-exposure to noise [52]. Road traffic is composed of light- and heavy-duty vehicles, with the emissions varying substantially between regions, depending on numerous factors such as the age of the fleet, used fuels, and the proportion of used technologies. Most of the internal combustion vehicles use diesel or gasoline as fuel. Compression-ignited diesel engines have the intrinsic feature of emitting more soot particles than spark-ignited gasoline, ethanol/methanol or gas-powered engines. The same applies to NO_x emissions, due to excess combustion air in compression-ignited engines [41]. However, diesel engine exhaust in traffic has already been regulated for a long time, especially in developed countries. Diesel particulate filters are highly effective in removing the soot particles from the exhaust [119], and later selective catalyst reduction has been effective in mitigating NO_x emissions [83, 86]. Therefore, in recent years, gasoline engines have become more significant emitters of particulate mass. In very recent regulations, the exhaust particle number concentrations have been limited, which has led to the increasing demand for adding particulate filters to gasoline exhaust systems [6], reducing significantly their particle number and black carbon emissions [98]. One major issue related to modern after-treatment technologies is, however, removing or tampering with these systems. Once they are removed, the emissions become directly representative of older technologies [31].

Depending on the region, alcohol or gas-powered engines can contribute a large share of the car fleet. Both, but especially gas-powered engines, usually have lower particulate mass emissions compared to gasoline [27, 98, 112]. However, also gaseous and semi-volatile emissions from traffic are potentially important, as they affect the secondary aerosol (SecA), or more specifically, secondary organic aerosol (SOA) formation in the atmosphere, which contributes increasingly to the air pollution mixture [34, 110, 112]. It can be seen that UFPs, semi-volatile organic compounds (SVOCs), and gaseous emissions play a crucial role in the future reduction of air pollution. However, there is significantly less research available on the health impacts of these emissions compared to the total ambient particulate mass of fine particles, i.e., ambient PM2.5. Nevertheless, concerning these traffic emissions and brain health, the neurotoxicity of traffic-originated SVOC emissions has been displayed in vitro [30]. Potentially, both UFPs and SVOCs may be especially harmful to brain health since these are probably able to penetrate bodily barriers and therefore cause effects also on brain health [8, 11]. When considering exposure to exhaust emissions, it is essential to recognize that the very small particulates, some just a few nanometres in size, as well as the effects of atmospheric aging and the formation of new particles from exhaust gases, have largely been overlooked in the past [38, 64].

Heavy-duty traffic, i.e. trucks and buses, contributes significantly to air quality and therefore to adverse health effects. Heavy-duty traffic has been almost exclusively diesel-powered, but the same emission regulations and after-treatment systems also apply to these [27]. Therefore, the emissions from heavy-duty traffic have been reduced similarly to light-duty traffic. In cities across the world, there are also increasing numbers of gas-powered and electric buses, which will further change the air quality in the cities. There is however, other heavy-duty equipment, including construction machinery and other non-road emissions that can contribute to local air quality. Although the current exhaust emission regulations for non-road vehicles follow the regulations for on-road vehicles, enforcing e.g. the use of DPF in new mobile machinery, the emissions from non-road mobile and stationary machinery have not been as strongly regulated as the on-road emissions, and together with the slow renewal of the fleet, the local exposure to these emissions may be significant and cause effects on human health [116].

The emissions from road traffic are not only from combustion engines. When combustion emissions decrease, other emissions become proportionally more important. Wear particles from brake discs and pads, road surfaces, and tires are much less studied than combustion particles [37]. While traffic is electrified, cars are becoming heavier due to the large batteries required for increased range. This can lead to higher emissions from road surfaces and tires [111]. However, the ability of electrical vehicles to capture energy via regenerative braking systems has been shown to result in decreased PM emissions from the brakes, yet some of the emissions will remain, especially from the tires, road surfaces and resuspension of the road dust [42]. Moreover, the metallic composition of brake discs and the composition of brake pads may significantly affect their toxicity. Since the metals are known contributors to brain diseases, further exploration concerning the topic seems particularly interesting [84, 85]. Compared to the combustion emissions, the mechanically produced particle emissions also result in different secondary aerosols, with considerably higher inorganic concentrations [111].

2.4.2 Marine Traffic

Marine traffic contributes significantly to air pollution-related health effects in human populations. Like other anthropogenic emissions sources, the emissions of ships and their characteristics can be significantly affected by technology choices [1, 108]. The atmospheric emissions from marine traffic are regulated by the International Maritime Organization (IMO) and by emission control areas (ECA), focusing especially on NO_x and sulphur emissions [32]. In marine vessels, particle emissions are not regulated, but it is known that they emit significant quantities of particles into the atmosphere [55], and the emissions are affected by fuel choices (e.g. low sulphur distilled fuels vs heavy fuel oil) and technologies used in gaseous emission reduction [56].

The sulphur content of the fuels used in marine traffic has been reduced, which directly affects the sulphate aerosol in the atmosphere [1]. More importantly, the fuels have shifted to lighter distillates, which affect the soot and ash emissions, which may have a larger positive effect on human health. However, the amount of fuel used for the marine sector is growing [48]. Heavy fuel oil is still allowed in the open seas, but it has less effect on human exposure. The emissions from the ships are usually aged in the atmosphere before people are exposed to them. Moreover, the local air quality in coastal areas is significantly affected by shipping emissions, port operations, and power generation in the ports. Therefore, the largest potential in reducing adverse health effects by the ships in ports is to avoid their emissions e.g. by using land electricity [32].

Recently, liquified natural gas has been introduced in new ships, significantly reducing particulate emissions (e.g. [58]). However, in all combustion emissions, there is still a risk for adverse health effects, partly because the particles emitted can be very small in size. Another disadvantage of natural gas-powered ships is the methane slip affecting the climate, which is considerable in lower engine loads [113]. Future maritime fuels also include ammonia, leading to a reduction of greenhouse gas emissions, but the possible health benefits from the change ultimately depend on the possible regulations and policies [122].

2.4.3 Aviation

Aviation is one of the combustion sources of traffic from which the emissions will be hard to eliminate. Aircraft turbines emit ultrafine soot particles and semi-volatile compounds that affect the air quality in the near vicinity of the airports due to aircraft operations, take-offs, and landings [61, 80]. High-altitude cruising cannot be considered to emit particles that would reach the human respiratory system since they are released above the mixing layer. However, the health-damaging potential of the emissions from ground operations and low-altitude emissions is significant at least near airports, and dependent on the operation frequency [7]. Indeed, there are measurement results that emissions from airports can affect the air quality for several kilometres [66]. In addition to nearby residential or commercial areas, the occupational exposure at the airports is potentially very high, and the workers are exposed to fresh emissions of UFPs from the aircraft engines and auxiliary power units.

2.5 Natural Sources of Air Pollution

2.5.1 *Wildfires*

Wildfires are becoming more common across the world. The factors affecting this phenomenon are the warming and drying of nature and the increasing population in the areas prone to wildfires [19]. Wildfires are uncontrolled combustion events that contribute to air pollution, and the composition of their emissions is difficult to predict. Overall, burning biomass emits high amounts of particulate and gaseous components that pose a risk to human health in large areas. The most exposed individuals are the members of firefighting crews working in the area of the fire, but these fires affect the air quality in large areas, and the exposed populations may be significantly large. Wildfire smoke contains a variety of potentially harmful compounds from the uncontrolled incomplete combustion of biological materials. They release vast amounts of carbon, unburned hydrocarbons, as well as combustion-derived metals and even soil components [13]. The smoke from wildfires is also contributing to the formation of secondary particles in the atmosphere. Wildfire smoke contains large amounts of compounds that could be harmful to human health and the brain, but the information is still sparse. Research indicates that emissions from wildfires carry greater toxicity and lead to more severe health impacts compared to many other sources [2, 53]. In vivo studies have also linked exposure to wildfire emissions to changes in gene expression in the prefrontal cortex of the brain [102]. Recently, a large cohort study found associations between exposure to wildfire emissions and dementia [24].

2.5.2 *Wind-Blown Dust, Road Dust, and Erosion*

Another concern for the air quality is coarse particles derived from soil. The deposition of these particles in the airways is very different from those derived from combustion processes, as is the composition of the particles. Wind-blown dust particles are usually quite large in size, and they are frequently released from arid areas across the world. Most likely, the larger particulate sizes do not impact brain health, at least directly. However, these particles can carry other compounds on their surfaces, which can then be released to the human upper respiratory tract after deposition. Indeed, the nasal area is one of the targets of inhalable particulates when considering the effects on the brain due to the short distance and physiological connection between the nasal cavity and the brain through olfactory nerves [51]. Therefore, it is not overruled that the soluble components on the surfaces of the coarse particles would affect the olfactory system. Moreover, it is known that dust particulates contain e.g. pesticides with long lifetimes that are neurotoxic agents [91, 107].

2.6 The Most Important Compositions for Brain Health Effects

In addition to the physical properties of the particles, their physicochemical composition is known to cause differing adverse effects on human health [50]. In the context of the brain, this has been less studied than the effects that have been studied longer, e.g. in respiratory and cardiac systems. Air pollution particles, such as UFPs, have a higher concentration of potentially harmful components concerning brain health due to increased surface area [127]. In many cases, the chemical components are condensed on the particle surface on top of the solid core. Alternatively, the UFPs may be condensed from volatile and semi-volatile compounds to particles in the atmosphere. The particles may have distinct physical and chemical features depending on the process.

2.6.1 Organic Compounds

Polycyclic aromatic hydrocarbons (PAHs) are a group of organic compounds absorbed onto the surface of particles. Their neurotoxic properties can be considered potentially harmful to brain health. For example, exposure to PAH emissions has been shown to induce degenerative changes in the brains of firefighters [54]. Especially, the developing brain can be susceptible to harmful effects. However, it is also known that PAH compounds are neurotoxic, and they can therefore be especially harmful to the brain. This can be seen as harmful, especially during the development stages of the brain [123]. Organic chemicals can quite easily distribute in human tissues and therefore cause effects beyond the exposure site, e.g., the lung [44]. In addition to PAH compounds, air pollution also contains other organic chemicals. Volatile and semi-volatile organic compounds can both form new particles and affect biological systems. These compounds are released from many combustion processes, e.g. engine exhaust, biomass, and fossil fuel combustion [68]. In addition, these compounds can be released from industrial processes, for example, petrochemical and solvent production, plastic industry, and building materials, causing indoor exposure [72]. Overall, organic compounds are also concerning in the context of brain health.

The organic compound group contains several substances known as forever chemicals, such as perfluoroalkyl and polyfluoroalkyl substances (PFAS). Recent studies have identified these chemicals as inhalation risks, highlighting their potential effects on respiratory health [22]. Moreover, PFASs may negatively affect the brain and cognition. They also accumulate in many brain regions, indicating the potential to cross the blood-brain barrier (BBB) [12]. In addition, many compounds used in plastic industries are known to accumulate in human tissues, and the environmental concentrations are increasing. These compounds are usually lipid soluble, which makes them ready to cross bodily barriers and reach the brain tissue. However, still rather little is known about the effects of these compounds, and therefore, more research would be needed to state the harmful effects on brain health [12].

2.6.2 *Metals and Inorganic Components*

A variety of metals are known to be neurotoxic and to affect brain health [100, 126]. In addition to the toxicity of the metals, many of them are needed in the normally functioning tissues. If the homeostasis of the metals is disturbed, e.g. by exposure to metals in air pollutants, it may cause functional changes in the tissues. Metals can be released in many combustion processes, especially from fossil fuels [69]. When released with the emissions, metallic compounds condense on the particle surfaces and can be transported in the human respiratory system. Depending on the solubility and oxidation state of the metals, they may differently affect the human tissues. Metal accumulation in the brain has been shown to negatively affect the brain and cause the onset of several neurological diseases [15, 70]. Negative associations of metallic compounds to neurodevelopment, motor performance, learning, and memory have been demonstrated [84, 85]. In addition to causing direct neurotoxicity, multiple metal elements have been linked to increased inflammation, which negatively impacts various systems in the human body [88].

The main inorganic gaseous pollutants from various combustion processes are nitrogen oxides and sulphur dioxide. These emissions also cause elevated ozone concentrations due to atmospheric processes [25]. Nitrogen oxides and ozone are oxidative compounds that can harm the brain by accelerating the biological aging process and impairing cognitive functions [17]. Instead, sulphur dioxide may have indirect effects on brain health by the potential to increase inflammation [124]. Eventually, these compounds form a major component of background aerosol, which, on the other hand, is not considered especially harmful.

2.7 Conclusions

Air pollution, especially particulate matter, but also gaseous pollutants, is undeniably a great hazard to human health. They are derived from various natural and anthropogenic sources, with different combustion processes being the most important. The source also determines the physicochemical properties of the released aerosol. When considering particulate matter, size is an important factor for adverse health effects. PM_{10} and $PM_{2.5}$ have been measured in the air for a long time, and there are legislation and limit values available for those pollutants. However, UFPs (<100 nm) are still largely unregulated. Their small size and large number concentration compared to the mass make them highly interesting from a brain health perspective. The size of the particles also determines their site of deposition and transport within and between tissues. The chemical composition of the pollutants is also a highly important factor for the adverse health effects.

Respiratory and cardiac diseases, as well as cancer, have been recognized to be caused by air pollutants for a long time. However, more evidence is accumulating on the harmfulness of air pollution to human brain health. Compared to cardiorespiratory diseases, the evidence on brain diseases is still sparser, and there is not much information on different air pollutant compositions that would be the most harmful to the brain. It is known that living near busy traffic increases the probability of e.g. neurodegenerative diseases. UFPs are one of the best candidates for effects on the central nervous system, since they can be effectively transported in human tissues, and can even reach the well-protected brain.

References

1. Aakko-Saksa, P. T., Lehtoranta, K., Kuittinen, N., Järvinen, A., Jalkanen, J.-P., Johnson, K., Jung, H., Ntziachristos, L., Gagné, S., Takahashi, C., Karjalainen, P., Rönkkö, T., & Timonen, H. (2023). Reduction in greenhouse gas and other emissions from ship engines: Current trends and future options. *Progress in Energy and Combustion Science, 94*, art. no. 101055. https://doi.org/10.1016/j.pecs.2022.101055
2. Aguilera, R., Corringham, T., Gershunov, A., et al. (2021). Wildfire smoke impacts respiratory health more than fine particles from other sources: Observational evidence from Southern California. *Nature Communications, 12*, 1493. https://doi.org/10.1038/s41467-021-21708-0
3. Alanen, J., Saukko, E., Lehtoranta, K., Murtonen, T., Timonen, H., Hillamo, R., Karjalainen, P., Kuuluvainen, H., Harra, J., Keskinen, J., & Rönkkö, T. (2015). The formation and physical properties of the particle emissions from a natural gas engine. *Fuel, 162*, 155–161.
4. Arnold, F., Pirjola, L., Rönkkö, T., Reichl, U., Schlager, H., Lähde, T., Heikkilä, J., & Keskinen, J. (2012). First on-line measurements of sulphuric acid gas in modern heavy duty diesel engine exhaust: Implications for nanoparticle formation. *Environmental Science and Technology, 46*, 11227–11234.
5. Aurela, M., Mylläri, F., Konist, A., Saarikoski, S., Olin, M., Simonen, P., Bloss, M., Nešumajev, D., Salo, L., Maasikmets, M., Sipilä, M., Maso, M. D., Keskinen, J., Timonen, H., & Rönkkö, T. (2021). Chemical and physical characterization of oil shale combustion emissions in Estonia. *Atmospheric Environment: X, 12*, art. no. 100139. https://doi.org/10.1016/j.aeaoa.2021.100139
6. Awad, O. I., Ma, X., Kamil, M., et al. (2020). Particulate emissions from gasoline direct injection engines: A review of how current emission regulations are being met by automobile manufacturers. *Science of the Total Environment, 718*, 137302. https://doi.org/10.1016/j.scitotenv.2020.137302

7. Bendtsen, K. M., Bengtsen, E., Saber, A. T., et al. (2021). A review of health effects associated with exposure to jet engine emissions in and around airports. *Environmental Health, 20*, 10. https://doi.org/10.1186/s12940-020-00690-y
8. Besis, A., Tsolakidou, A., Balla, D., Samara, C., Voutsa, D., Pantazaki, A., Choli-Papadopoulou, T., & Lialiaris, T. S. (2017). Toxic organic substances and marker compounds in size segregated urban particulate matter—Implications for involvement in the in vitro bioactivity of the extractable organic matter. *Environmental Pollution, 230*, 758–774. https://doi.org/10.1016/j.envpol.2017.06.096
9. Bongaerts, E., Lecante, L. L., Bové, H., Roeffaers, M. B., Ameloot, M., Fowler, P. A., & Nawrot, T. S. (2022). Maternal exposure to ambient black carbon particles and their presence in maternal and fetal circulation and organs: An analysis of two independent population-based observational studies. *The Lancet Planetary Health, 6*(10), e804–e811. https://doi.org/10.1016/S2542-5196(22)00200-5
10. Calderón-Garcidueñas, L., & Ayala, A. (2022). Air pollution, ultrafine particles, and your brain: Are combustion nanoparticle emissions and engineered nanoparticles causing preventable fatal neurodegenerative diseases and common neuropsychiatric outcomes? *Environmental Science & Technology, 56*(11), 6847–6856. https://doi.org/10.1021/acs.est.1c04706
11. Calderón-Garcidueñas, L., Reynoso-Robles, R., & González-Maciel, A. (2019). Combustion and friction-derived nanoparticles and industrial-sourced nanoparticles: The culprit of Alzheimer and Parkinson's diseases. *Environmental Research, 176*, 108574. https://doi.org/10.1016/j.envres.2019.108574
12. Cao, Y., & Ng, C. (2021). Absorption, distribution, and toxicity of per-and polyfluoroalkyl substances (PFAS) in the brain: A review. *Environmental Science: Processes & Impacts, 23*(11), 1623–1640. https://doi.org/10.1039/D1EM00228G
13. Cascio, W. E. (2018). Wildland fire smoke and human health. *Science of the Total Environment, 624*, 586–595. https://doi.org/10.1016/j.scitotenv.2017.12.086
14. Cassee, F., Morawska, L., Peters, A., Wierzbicka, A., Buonanno, G., Cyrys, J., Schnelle-Kreis, J., Kowalski, M., Riediker, M., Birmili, W., Querol, X., Yildirim, A., Elder, A., Yu, I. J., Ovrevik, J., Hougaard, K., Loft, S., Schmid, O., Stoeger, T., & Lucht, S. (2019). *Ambient ultrafine particles: Evidence for policy makers*. A report prepared by the 'Thinking outside the box' team [WWW Document]. https://efca.net/files/WHITE%20PAPER-UFP%20evidence%20for%20policy%20makers%20(25%20OCT).pdf. Accessed 3 July 2023.
15. Chen, P., Miah, M. R., & Aschner, M. (2016). Metals and neurodegeneration. *F1000Research, 5*, F1000 Faculty Rev-366. https://doi.org/10.12688/f1000research.7431.1
16. Chen, H., Kwong, J. C., Copes, R., Tu, K., Villeneuve, P. J., van Donkelaar, A., Hystad, P., Martin, R. V., Murray, B. J., Jessiman, B., Wilton, A. S., Kopp, A., & Burnett, R. T. (2017). Living near major roads and the incidence of dementia, Parkinson's disease, and multiple sclerosis: A population-based cohort study. *Lancet, 389*(10070), 718–726. https://doi.org/10.1016/S0140-6736(16)32399-6
17. Chen, Y., Dong, Y., Zhang, Y., Xia, D., Wang, Y., Wang, Y., Cai, Y., & Hu, F. (2024). Effects of major air pollutants on cognitive function in middle-aged and elderly adults: Panel data evidence from China Health and Retirement Longitudinal Study. *Journal of Global Health, 14*, 04153. https://doi.org/10.7189/jogh.14.04153
18. Cordell, R. L., Mazet, M., Dechoux, C., Hama, S. M. L., Staelens, J., Hofman, J., et al. (2016). Evaluation of biomass burning across North West Europe and its impact on air quality. *Atmospheric Environment, 141*, 276–286. https://doi.org/10.1016/j.atmosenv.2016.06.065
19. Cunningham, C. X., Williamson, G. J., & Bowman, D. M. J. S. (2024). Increasing frequency and intensity of the most extreme wildfires on Earth. *Nature Ecology & Evolution, 8*, 1420–1425. https://doi.org/10.1038/s41559-024-02452-2
20. Damayanti, S., Harrison, R. M., Pope, F., & Beddows, D. C. S. (2023). Limited impact of diesel particle filters on road traffic emissions of ultrafine particles. *Environment International, 174*, art. no. 107888. https://doi.org/10.1016/j.envint.2023.107888

21. De Filippo, A., & Maricq, M. M. (2008). Diesel nucleation mode particles: Semivolatile or solid? *Environmental Science and Technology, 42*(21), 7957–7962. https://doi.org/10.1021/es8010332
22. De Silva, A. O., Armitage, J. M., Bruton, T. A., Dassuncao, C., Heiger-Bernays, W., Hu, X. C., Kärrman, A., Kelly, B., Ng, C., Robuck, A., Sun, M., Webster, T. F., & Sunderland, E. M. (2021). PFAS exposure pathways for humans and wildlife: A synthesis of current knowledge and key gaps in understanding. *Environmental Toxicology and Chemistry, 40*, 631–657. https://doi.org/10.1002/etc.4935
23. Elder, A., Gelein, R., Silva, V., Feikert, T., Opanashuk, L., Carter, J., et al. (2006). Translocation of inhaled ultrafine manganese oxide particles to the central nervous system. *Environmental Health Perspectives, 114*(8), 1172–1178. https://doi.org/10.1289/ehp.9030
24. Elser, H., Frankland, T. B., Chen, C., et al. (2025). Wildfire smoke exposure and incident dementia. *JAMA Neurology, 82*(1), 40–48. https://doi.org/10.1001/jamaneurol.2024.4058
25. Erickson, L. E., Newmark, G. L., Higgins, M. J., & Wang, Z. (2020). Nitrogen oxides and ozone in urban air: A review of 50 plus years of progress. *Environmental Progress & Sustainable Energy, 39*, e13484. https://doi.org/10.1002/ep.13484
26. European Commission, 'Directive (EU) 2024/2881 of the European Parliament and of the Council of 23 October 2024 on ambient air quality and cleaner air for Europe recast' PE/88/2024/REV/1.
27. European Environment Agency. (2022). *Air quality e-reporting (AQ e-Reporting).* https://sdi.eea.europa.eu/catalogue/srv/api/records/3b390c9c-f321-490a-b25aae93b2ed80c1
28. European Environment Agency (EEA). (2022). *Transport and environment report 2021. Decarbonising road transport—The role of vehicles, fuels and transport demand.* 87 pp. ISBN: 978-92-9480-473-0. https://doi.org/10.2800/68902
29. Garcia, G. J., Schroeter, J. D., & Kimbell, J. S. (2015). Olfactory deposition of inhaled nanoparticles in humans. *Inhalation Toxicology, 27*(8), 394–403. https://doi.org/10.3109/08958378.2015.1066904
30. Gerber, L.-S., de Leijer, D. C. A., Rujas Arranz, A., Lehmann, J. M. M. L., & Verheul, M. E. (2024). Comparison of the neurotoxic potency of different ultrafine particle fractions from diesel en-gine exhaust following direct and simulated inhalation exposure. *The Science of the Total Environment., 951*, 175469. https://doi.org/10.1016/j.scitotenv.2024.175469
31. Giechaskiel, B., Forloni, F., Carriero, M., Baldini, G., Castellano, P., Vermeulen, R., Kontses, D., Fragkiadoulakis, P., Samaras, Z., & Fontaras, G. (2022). Effect of tampering on on-road and off-road diesel vehicle emissions. *Sustainability, 14*(10), 6065. https://doi.org/10.3390/su14106065
32. Gössling, S., Meyer-Habighorst, C., & Humpe, A. (2021). A global review of marine air pollution policies, their scope and effectiveness. *Ocean and Coastal Management, 212*, 105824. https://doi.org/10.1016/j.ocecoaman.2021.105824
33. Hakkarainen, H., Salo, L., Mikkonen, S., et al. (2022). Black carbon toxicity dependence on particle coating: Measurements with a novel cell exposure method. *Science of the Total Environment, 838*(Part 4), 156543., ISSN 0048-9697. https://doi.org/10.1016/j.scitotenv.2022.156543
34. Hallquist, M., Wenger, J. C., Baltensperger, U., Rudich, Y., Simpson, D., Claeys, M., Dommen, J., Donahue, N. M., George, C., Goldstein, A. H., Hamilton, J. F., Herrmann, H., Hoffmann, T., Iinuma, Y., Jang, M., Jenkin, M. E., Jimenez, J. L., Kiendler-Scharr, A., Maenhaut, W., McFiggans, G., Mentel, T. F., Monod, A., Prévôt, A. S. H., Seinfeld, J. H., Surratt, J. D., Szmigielski, R., & Wildt, J. (2009). The formation, properties and impact of secondary organic aerosol: Current and emerging issues. *Atmospheric Chemistry and Physics, 9*(14), 5155–5236. https://doi.org/10.5194/acp-9-5155-2009
35. Han, D., Chen, R., Kan, H., & Xu, Y. (2023). The bio-distribution, clearance pathways, and toxicity mechanisms of ambient ultrafine particles. *Eco-Environment & Health, 2*, 95–106. https://doi.org/10.1016/j.eehl.2023.06.001

36. Happonen, M., Mylläri, F., Karjalainen, P., Frey, A., Saarikoski, S., Carbone, S., Hillamo, R., Pirjola, L., Häyrinen, A., Kytömäki, J., Niemi, J., Keskinen, J., & Rönkkö, T. (2013). Size distribution, chemical composition, and hygroscopicity of fine particles emitted from an oil-fired heating plant. *Environmental Science and Technology, 47*, 14468–14475.
37. Harrison, R. M., Allan, J., Carruthers, D., Heal, M. R., Lewis, A. C., Marner, B., Murrells, T., & Williams, A. (2021). Non-exhaust vehicle emissions of particulate matter and VOC from road traffic: A review. *Atmospheric Environment, 262*, 118592. https://doi.org/10.1016/j.atmosenv.2021.118592
38. Hartikainen, A. H., Ihalainen, M., Yli-Pirilä, P., Hao, L., Kortelainen, M., Pieber, S. M., & Sippula, O. (2023). Photochemical transformation and secondary aerosol formation potential of Euro6 gasoline and diesel passenger car exhaust emissions. *Journal of Aerosol Science, 171*, 106159. https://doi.org/10.1016/j.jaerosci.2023.106159
39. Health Effects Institute (HEI). (2013). *Understanding the health effects of ambient ultrafine particles*. Health Effects Institute.
40. Heusinkveld, H., Wahle, T., Campbell, A., et al. (2016). Neurodegenerative and neurological disorders by small inhaled particles. *Neurotoxicology, 113*, 94–106. https://doi.org/10.1016/j.neuro.2016.07.007
41. Heywood, J. B. (2018). *Internal combustion engine fundamentals* (2nd ed.). McGraw-Hill Education.
42. Hicks, W., Green, D. C., & Beevers, S. (2023). Quantifying the change of brake wear particulate matter emissions through powertrain electrification in passenger vehicles. *Environmental Pollution, 336*, 122400. https://doi.org/10.1016/j.envpol.2023.122400
43. Hietikko, R., Kuuluvainen, H., Harrison, R. M., Portin, H., Timonen, H., Niemi, J. V., & Rönkkö, T. (2018). Diurnal variation of nanocluster aerosol concentrations and emission factors in a street canyon. *Atmospheric Environment, 189*, 98–106. https://doi.org/10.1016/j.atmosenv.2018.06.031
44. Holme, J. A., Brinchmann, B. C., Refsnes, M., Låg, M., & Øvrevik, J. (2019). Potential role of polycyclic aromatic hydrocarbons as mediators of cardiovascular effects from combustion particles. *Environmental Health, 18*(1), 74. https://doi.org/10.1186/s12940-019-0514-2
45. Hopkins, L. E., Laing, E. A., Peake, J. L., Uyeminami, D., Mack, S. M., Li, X., et al. (2018). Repeated iron–soot exposure and nose-to-brain transport of inhaled ultrafine particles. *Toxicologic Pathology, 46*(1), 75–84. https://doi.org/10.1177/0192623317729222
46. Hovorka, J., Pokorná, P., Hopke, P. K., Křůmal, K., Mikuška, P., & Píšová, M. (2015). Wood combustion, a dominant source of winter aerosol in residential district in proximity to a large automobile factory in Central Europe. *Atmospheric Environment, 113*, 98–107. https://doi.org/10.1016/j.atmosenv.2015.04.068
47. IEA. (2021). *Global fuel economy initiate 2021, vehicle fuel economy in major markets 2000–2019*. https://www.iea.org/reports/global-fuel-economy-initiative-2021
48. IEA (International Energy Agency). (2020). *International shipping*. Available: https://www.iea.org/reports/international-shipping
49. International Energy Agency (IEA). (2016). *Weo-2016 special report energy and air pollution* (p. 266). International Energy Agency.
50. Jalava, P. I., Wang, Q., Kuuspalo, K., Ruusunen, J., Hao, L., Fang, D., & Hirvonen, M.-R. (2015). Day and night variation in chemical composition and toxicological responses of size segregated urban air PM samples in a high air pollution situation. *Atmospheric Environment, 120*, 427–437. https://doi.org/10.1016/j.atmosenv.2015.08.089
51. Kanninen, K. M., Lampinen, R., Rantanen, L. M., Odendaal, L., Jalava, P., Chew, S., & White, A. R. (2020). Olfactory cell cultures to investigate health effects of air pollution exposure: Implications for neurodegeneration. *Neurochemistry International, 136*, 104729. https://doi.org/10.1016/j.neuint.2020.104729
52. Khan, J., Ketzel, M., Kakosimos, K., Sørensen, M., & Jensen, S. S. (2018). Road traffic air and noise pollution exposure assessment—A review of tools and techniques. *Science of the Total Environment, 634*, 661–676. https://doi.org/10.1016/j.scitotenv.2018.03.374

53. Kim, Y. H., Warren, S. H., Krantz, Q. T., King, C., Jaskot, R., Preston, W. T., et al. (2018). Mutagenicity and lung toxicity of smoldering vs. flaming emissions from various biomass fuels: Implications for health effects from wildland fires. *Environmental Health Perspectives, 126*(1), 017011. https://doi.org/10.1289/EHP2200
54. Kim, Y. T., Kim, W., Bae, M., Choi, J. E., Kim, M.-J., Oh, S. S., Park, K. S., Park, S., Lee, S.-K., Koh, S.-B., & Kim, C. (2022). The effect of polycyclic aromatic hydrocarbons on changes in the brain structure of firefighters: An analysis using data from the Firefighters Research on Enhancement of Safety & Health study. *Science of the Total Environment, 816*, 151655. https://doi.org/10.1016/j.scitotenv.2021.151655
55. Kuittinen, N., Jalkanen, J.-P., Alanen, J., Ntziachristos, L., Hannuniemi, H., Johansson, L., Karjalainen, P., Saukko, E., Isotalo, M., Aakko-Saksa, P., Lehtoranta, K., Keskinen, J., Simonen, P., Saarikoski, S., Asmi, E., Laurila, T., Hillamo, R., Mylläri, F., Lihavainen, H., Timonen, H., & Rönkkö, T. (2021). Shipping remains a globally significant source of anthropogenic PN emissions even after 2020 sulfur regulation. *Environmental Science and Technology, 55*(1), 129–138. https://doi.org/10.1021/acs.est.0c03627
56. Kuittinen, N., Timonen, H., Karjalainen, P., Murtonen, T., Vesala, H., Bloss, M., Honkanen, M., Lehtoranta, K., Aakko-Saksa, P., & Rönkkö, T. (2024). In-depth characterization of exhaust particles performed on-board a modern cruise ship applying a scrubber. *Science of the Total Environment, 946*, art. no. 174052. https://doi.org/10.1016/j.scitotenv.2024.174052
57. Kwon, H. S., Ryu, M. H., & Carlsten, C. (2020). Ultrafine particles: unique physicochemical properties relevant to health and disease. *Experimental & Molecular Medicine, 52*, 318–328. https://doi.org/10.1038/s12276-020-0405-1
58. Lehtoranta, K., Aakko-Saksa, P., Murtonen, T., Vesala, H., Ntziachristos, L., Rönkkö, T., Karjalainen, P., Kuittinen, N., & Timonen, H. (2019). Particulate mass and nonvolatile particle number emissions from marine engines using low-sulfur fuels, natural gas, or scrubbers. *Environmental Science and Technology, 53*(6), 3315–3322. https://doi.org/10.1021/acs.est.8b05555
59. Lelieveld, J., Haines, A., Burnett, R., Tonne, C., Klingmüller, K., Münzel, T., et al. (2023). Air pollution deaths attributable to fossil fuels: Observational and modelling study. *BMJ, 383*, e077784. https://doi.org/10.1136/bmj-2023-077784
60. Lepistö, T., Kuuluvainen, H., Juuti, P., Järvinen, A., Arffman, A., & Rönkkö, T. (2020). Measurement of the human respiratory tract deposited surface area of particles with an electrical low-pressure impactor. *Aerosol Science and Technology, 54*(8), 958–971. https://doi.org/10.1080/02786826.2020.1745141
61. Lepistö, T., Barreira, L. M. F., Helin, A., Niemi, J. V., Kuittinen, N., Lintusaari, H., Silvonen, V., Markkula, L., Manninen, H. E., Timonen, H., Jalava, P., Saarikoski, S., & Rönkkö, T. (2023). Snapshots of wintertime urban aerosol characteristics: Local sources emphasized in ultrafine particle number and lung deposited surface area. *Environmental Research, 231*, art. no. 116068. https://doi.org/10.1016/j.envres.2023.116068
62. Li, J., Li, J., Wang, G., Ho, K. F., Dai, W., Zhang, T., et al. (2021). Effects of atmospheric aging processes on in vitro induced oxidative stress and chemical composition of biomass burning aerosols. *Journal of Hazardous Materials, 401*, 123750. https://doi.org/10.1016/j.jhazmat.2020.123750
63. Lin, W., Dai, J., Liu, R., Zhai, Y., Yue, D., & Hu, Q. (2019). Integrated assessment of health risk and climate effects of black carbon in the Pearl River Delta region, China. *Environmental Research, 176*, 2019. https://doi.org/10.1016/j.envres.2019.06.003
64. Lintusaari, H., Kuuluvainen, H., Vanhanen, J., Salo, L., Portin, H., Järvinen, A., Juuti, P., Hietikko, R., Teinilä, K., Timonen, H., Niemi, J. V., & Rönkkö, T. (2023). Sub-23 nm particles dominate non-volatile particle number emissions of road traffic. *Environmental Science & Technology, 57*(29), 10763–10772. https://doi.org/10.1021/acs.est.3c03221
65. Lintusaari, H., Lepistö, T., Saarikoski, S., Salo, L., Silvonen, V., Barreira, L. M. F., Aurela, M., Hoivala, J., Markkula, L., Ondracek, J., Wahle, T., Vojtisek-Lom, M., Topinka, J., Schins, R. P. F., Jalava, P., Timonen, H., Kanninen, K. M., & Rönkkö, T. (2025). Importance of

sub-23 nm particles in traffic environments: Particle number emission factors and extra-thoracic deposition doses. *Environmental Pollution, 369*, 125835. https://doi.org/10.1016/j.envpol.2025.125835
66. Lopes, M., Russo, A., Monjardino, J., Gouveia, C., & Ferreira, F. (2019). Monitoring of ultrafine particles in the surrounding urban area of a civilian airport. *Atmospheric Pollution Research, 10*(5), 1454–1463. https://doi.org/10.1016/j.apr.2019.04.002
67. Luoma, K., Niemi, J. V., Aurela, M., Lun Fung, P., Helin, A., Hussein, T., Kangas, L., Kousa, A., Rönkkö, T., Timonen, H., Virkkula, A., & Petäjä, T. (2021). Spatiotemporal variation and trends in equivalent black carbon in the Helsinki metropolitan area in Finland. *Atmospheric Chemistry and Physics, 21*(2) art. no. 59, 1173–1189. https://doi.org/10.5194/acp-21-1173-2021
68. Ma, S., Yue, C., Tang, J., Lin, M., Zhuo, M., et al. (2021). Occurrence and distribution of typical semi-volatile organic chemicals (SVOCs) in paired indoor and outdoor atmospheric fine particle samples from cities in southern China. *Environmental Pollution (1987), 269*, 116123. https://doi.org/10.1016/j.envpol.2020.116123
69. Maciejczyk, P., Chen, L.-C., & Thurston, G. (2021). The role of fossil fuel combustion metals in PM2.5 air pollution health associations. *Atmosphere, 12*(9), 1086. https://doi.org/10.3390/atmos12091086
70. Maher, B. A., Ahmed, I. A., Karloukovski, V., MacLaren, D. A., Foulds, P. G., Allsop, D., et al. (2016). Magnetite pollution nanoparticles in the human brain. *Proceedings of the National Academy of Sciences, 113*(39), 10797–10801. https://doi.org/10.1073/pnas.1605941113
71. Mastral, A. M., Callen, M. S., & Garcia, T. (2000). Toxic organic emissions from coal combustion. *Fuel Processing Technology, 67*(1), 1–10. https://doi.org/10.1016/S0378-3820(00)00088-6
72. McDonald, B. C., et al. (2018). Volatile chemical products emerging as largest petrochemical source of urban organic emissions. *Science, 359*, 760–764. https://doi.org/10.1126/science.aaq0524
73. Milani, C., Farina, F., Botto, L., Massimino, L., Lonati, E., Donzelli, E., Ballarini, E., Crippa, L., Marmiroli, P., Bulbarelli, A., & Palestini, P. (2020). Systemic exposure to air pollution induces oxidative stress and inflammation in mouse brain, contributing to neurodegeneration onset. *International Journal of Molecular Sciences, 21*(10), 3699. https://doi.org/10.3390/ijms21103699
74. Molina, C., Toro, A. R., Manzano, C. A., Canepari, S., Massimi, L., & Leiva-Guzmán, M. A. (2020). Airborne aerosols and human health: Leapfrogging from mass concentration to oxidative potential. *Atmosphere, 11*(9), 917. https://doi.org/10.3390/atmos11090917
75. Mylläri, F., Asmi, E., Anttila, T., Saukko, E., Vakkari, V., Pirjola, L., Hillamo, R., Laurila, T., Häyrinen, A., Rautiainen, J., Lihavainen, H., O'Connor, E., Niemelä, V., Keskinen, J., Dal Maso, M., & Rönkkö, T. (2016). New particle formation in the fresh flue-gas plume from a coal-fired power plant: Effect of flue-gas cleaning. *Atmospheric Chemistry and Physics, 16*, 7485–7496. https://doi.org/10.5194/acp-16-7485-2016
76. Nemmar, A., Hoet, P. H., Vanquickenborne, B., Dinsdale, D., Thomeer, M., Hoylaerts, M. F., Vanbilloen, H., Mortelmans, L., & Nemery, B. (2002). Passage of inhaled particles into the blood circulation in humans. *Circulation, 105*(4), 411–414. https://doi.org/10.1161/hc0402.104118
77. Nosko, O., Vanhanen, J., & Olofsson, U. (2017). Emission of 1.3–10 nm airborne particles from brake materials. *Aerosol Science and Technology, 51*(1), 91–96. https://doi.org/10.1080/02786826.2016.1255713
78. Oberdörster, G., Sharp, Z., Atudorei, V., Elder, A., Gelein, R., Lunts, A., Kreyling, W., & Cox, C. (2002). Extrapulmonary translocation of ultrafine carbon particles following whole-body inhalation exposure of rats. *Journal of Toxicology and Environmental Health. Part A, 65*, 1531–1543.
79. Olin, M., Kuuluvainen, H., Aurela, M., Kalliokoski, J., Kuittinen, N., Isotalo, M., Timonen, H. J., Niemi, J. V., Rönkkö, T., & Dal Maso, M. (2019). Traffic-originated nanocluster

emission exceeds H2SO4-driven photochemical new particle formation in an urban area. *Atmospheric Chemistry and Physics, 20*(1), 1–13. https://doi.org/10.5194/acp-20-1-2020
80. Owen, B., Anet, J. G., Bertier, N., Christie, S., Cremaschi, M., Dellaert, S., Edebeli, J., Janicke, U., Kuenen, J., Lim, L., et al. (2022). Review: Particulate matter emissions from aircraft. *Atmosphere, 13*(8), 1230. https://doi.org/10.3390/atmos13081230
81. Peters, R., Ee, N., Peters, J., et al. (2019). Air pollution and dementia: A systematic review. *Journal of Alzheimer's Disease, 70*(s1), S145–S163. https://doi.org/10.3233/JAD-180631
82. Pirjola, L., Karl, M., Rönkkö, T., & Arnold, F. (2015). Model studies of volatile diesel exhaust particle formation: Are organic vapours involved in nucleation and growth? *Atmospheric Chemistry and Physics, 15*, 10435–10452. https://doi.org/10.5194/acp-15-10435-2015
83. Platt, S. M., El Haddad, I., Pieber, S. M., et al. (2017). Gasoline cars produce more carbonaceous particulate matter than modern filter-equipped diesel cars. *Scientific Reports, 7*(1), 4926–4929. https://doi.org/10.1038/s41598-017-03714-9
84. Potter, N. A., Meltzer, G. Y., Avenbuan, O. N., Raja, A., & Zelikoff, J. T. (2021a). Particulate matter and associated metals: A link with neurotoxicity and mental health. *Atmosphere (Basel)., 12*(4), 425. https://doi.org/10.3390/atmos12040425
85. Potter, N. A., Meltzer, G. Y., Avenbuan, O. N., Raja, A., & Zelikoff, J. T. (2021b). Particulate matter and associated metals: A link with neurotoxicity and mental health. *Atmosphere, 12*, 425. https://doi.org/10.3390/atmos12040425
86. Praveeno, V., & Martin, M. (2018). A review of various treatment techniques to reduce NOx in CI engines. *Journal of the Energy Institute, 91*, 704–720. https://doi.org/10.1016/j.joei.2017.05.010
87. Presto, A. A., Saha, P. K., & Robinson, A. L. (2021). Past, present, and future of ultrafine particle exposures in North America. *Atmospheric Environment: X, 10*, 100109. https://doi.org/10.1016/j.aeaoa.2021.100109
88. Qiao, H., Liu, W., Gu, H., Wang, D., & Wang, Y. (2015). The transport and deposition of nanoparticles in respiratory system by inhalation. *Journal of Nanomaterials, 2015*, 394507, 8 pages. https://doi.org/10.1155/2015/394507
89. Rentschler, J., & Leonova, N. (2023). Global air pollution exposure and poverty. *Nature Communications, 14*(1), 4432. https://doi.org/10.1038/s41467-023-39797-4
90. Ribeiro, C. B., Rodella, F. H. C., & Hoinaski, L. (2022). Regulating light-duty vehicle emissions: An overview of US, EU, China and Brazil programs and its effect on air quality. *Clean Technologies and Environmental Policy, 24*, 851–862. https://doi.org/10.1007/s10098-021-02238-1
91. Richardson, J. R., Fitsanakis, V., Westerink, R. H., & Kanthasamy, A. G. (2019). Neurotoxicity of pesticides. *Acta Neuropathologica, 138*, 343–362.
92. Rönkkö, T., Virtanen, A., Kannosto, J., Keskinen, J., Lappi, M., & Pirjola, L. (2007). Nucleation mode particles with a nonvolatile core in the exhaust of a heavy-duty diesel vehicle. *Environmental Science and Technology, 41*, 6384–6389.
93. Rönkkö, T., Pirjola, L., Ntziachristos, L., Heikkilä, J., Karjalainen, P., Hillamo, R., & Keskinen, J. (2014). Vehicle engines produce exhaust nanoparticles even when not fueled. *Environmental Science and Technology, 48*, 2043–2050.
94. Rönkkö, T., Kuuluvainen, H., Karjalainen, P., Keskinen, J., Hillamo, R., Niemi, J. V., Pirjola, L., Timonen, H. J., Saarikoski, S., Saukko, E., Järvinen, A., Silvennoinen, H., Rostedt, A., Olin, M., Yli-Ojanperä, J., Nousiainen, P., Kousa, A., & Dal Maso, M. (2017). Traffic is a major source of atmospheric nanocluster aerosol. *Proceedings of the National Academy of Sciences of the United States of America (PNAS), 114*(29), 7549–7554. https://doi.org/10.1073/pnas.1700830114
95. Rönkkö, T., Saarikoski, S., Kuittinen, N., Karjalainen, P., Keskinen, H., Järvinen, A., Mylläri, F., Aakko-Saksa, P., & Timonen, H. (2023). Review of black carbon emission factors from different anthropogenic sources. *Environmental Research Letters, 18*, 033004. https://doi.org/10.1088/1748-9326/acbb1b

96. Rosário Filho, N. A., Urrutia-Pereira, M., d'Amato, G., Cecchi, L., Ansotegui, I. J., Galán, C., et al. (2021). Air pollution and indoor settings. *World Allergy Organization Journal, 14*(1), 100499. https://doi.org/10.1016/j.waojou.2020.100499
97. Saarikoski, S., Timonen, H., Saarnio, K., Aurela, M., Järvi, L., Keronen, P., Kerminen, V.-M., & Hillamo, R. (2008). Sources of organic carbon in fine particulate matter in northern European urban air. *Atmospheric Chemistry and Physics, 8*, 6281–6295. https://doi.org/10.5194/acp-8-6281-2008
98. Saarikoski, S., Järvinen, A., Markkula, L., Aurela, M., Kuittinen, N., Hoivala, J., Barreira, L. M. F., Aakko-Saksa, P., Lepistö, T., Marjanen, P., Timonen, H., Hakkarainen, H., Jalava, P., & Rönkkö, T. (2024). Towards zero pollution vehicles by advanced fuels and exhaust aftertreatment technologies. *Environmental Pollution, 347*, art. no. 123665. https://doi.org/10.1016/j.envpol.2024.123665
99. Saarnio, K., Frey, A., Niemi, J. V., Timonen, H., Rönkkö, T., Karjalainen, P., Vestenius, M., Teinilä, K., Pirjola, L., Niemelä, V., Keskinen, J., Häyrinen, A., & Hillamo, R. (2014). Chemical composition and size of particles in emissions of a coal-fired power plant with flue gas desulfurization. *Journal of Aerosol Science, 73*, 14–26.
100. Sanders, T., Liu, Y., Buchner, V., & Tchounwou, P. B. (2009). Neurotoxic effects and biomarkers of lead exposure: A review. *Reviews on Environmental Health, 24*(1), 15–46. https://doi.org/10.1515/REVEH.2009.24.1.15
101. Schraufnagel, D. E. (2020). The health effects of ultrafine particles. *Experimental & Molecular Medicine, 52*, 311–317. https://doi.org/10.1038/s12276-020-0403-3
102. Schuller, A., Oakes, J., LaRocca, T., Matz, J., Eden, M., Bellini, C., & Luke Montrose, L. (2024). Robust differential gene expression patterns in the prefrontal cortex of male mice exposed to an occupationally relevant dose of laboratory-generated wildfire smoke. *Toxicological Sciences, 201*(2), 300–310. https://doi.org/10.1093/toxsci/kfae097
103. Sgro, L. A., Sementa, P., Vaglieco, B. M., Rusciano, G., D'Anna, A., & Minutolo, P. (2012). Investigating the origin of nuclei particles in GDI engine exhausts. *Combustion and Flame, 159*(4), 1687–1692. https://doi.org/10.1016/j.combustflame.2011.12.013
104. Shaddick, G., Thomas, M. L., Mudu, P., et al. (2020). Half the world's population are exposed to increasing air pollution. *npj Climate and Atmospheric Science, 3*, 23. https://doi.org/10.1038/s41612-020-0124-2
105. Silvonen, V., Salo, L., Raunima, T., Vojtisek-Lom, M., Ondracek, J., Topinka, J., Schins, R. P. F., Lepistö, T., Lintusaari, H., Saarikoski, S., Barreira, L. M. F., Hoivala, J., Markkula, L., Kulmala, I., Vinha, J., Karjalainen, P., & Rönkkö, T. (2023). Lung-depositing surface area (LDSA) of particles in office spaces around Europe: Size distributions, I/O-ratios and infiltration. *Building and Environment, 246*, art. no. 110999. https://doi.org/10.1016/j.buildenv.2023.110999
106. Sipilä, M., Berndt, T., Petäjä, T., Brus, D., Vanhanen, J., Stratmann, F., Patokoski, J., Mauldin, R. L., III, Hyvärinen, A.-P., Lihavainen, H., & Kulmala, M. (2010). The role of sulfuric acid in atmospheric nucleation. *Science, 327*(5970), 1243–1246. https://doi.org/10.1126/science.1180315
107. Socorro, J., Durand, A., Temime-Roussel, B., et al. (2016). The persistence of pesticides in atmospheric particulate phase: An emerging air quality issue. *Scientific Reports, 6*, 33456. https://doi.org/10.1038/srep33456
108. Sofiev, M., Winebrake, J. J., Johansson, L., Carr, E. W., Prank, M., Soares, J., Vira, J., Kouznetsov, R., Jalkanen, J., & Corbett, J. J. (2018). Cleaner fuels for ships provide public health benefits with climate tradeoffs. *Nature Communications, 9*, 406.
109. Song, X., Hu, Y., Ma, Y., et al. (2022). Is short-term and long-term exposure to black carbon associated with cardiovascular and respiratory diseases? A systematic review and meta-analysis based on evidence reliability. *BMJ Open, 12*, e049516. https://doi.org/10.1136/bmjopen-2021-049516

110. Srivastava, D., Vu, T. V., Tong, S., Shi, Z., & Harrison, R. (2022). Formation of secondary organic aerosols from anthropogenic precursors in laboratory studies. *npj Climate and Atmospheric Science, 5*, 22. https://doi.org/10.1038/s41612-022-00238-6
111. Timmers, V. R. J. H., & Achten, P. A. J. (2016). Non-exhaust PM emissions from electric vehicles. *Atmospheric Environment, 134*, 10–17. https://doi.org/10.1016/j.atmosenv.2016.03.017
112. Timonen, H., Karjalainen, P., Saukko, E., Saarikoski, S., Aakko-Saksa, P., Simonen, P., Murtonen, T., Dal Maso, M., Kuuluvainen, H., Bloss, M., Ahlberg, E., Svenningsson, B., Pagels, J., Brune, W. H., Keskinen, J., Worsnop, D. R., Hillamo, R., & Rönkkö, T. (2017). Influence of fuel ethanol content on primary emissions and secondary aerosol formation potential for a modern flex-fuel gasoline vehicle. *Atmospheric Chemistry and Physics, 17*, 5311–5329. https://doi.org/10.5194/acp-17-5311-2017
113. Ushakov, S., Stenersen, D., & Einang, P. M. (2019). Methane slip from gas fuelled ships: A comprehensive summary based on measurement data. *Journal of Marine Science and Technology, 24*, 1308–1325. https://doi.org/10.1007/s00773-018-00622-z
114. Vanbrabant, K., Van Dam, D., Bongaerts, E., et al. (2024). Accumulation of ambient black carbon particles within key memory-related brain regions. *JAMA Network Open, 7*(4), e245678. https://doi.org/10.1001/jamanetworkopen.2024.5678
115. Vicente, E. D., & Alves, C. A. (2018). An overview of particulate emissions from residential biomass combustion. *Atmospheric Research, 199*, 159–185. https://doi.org/10.1016/j.atmosres.2017.08.027
116. Waluś, K. J., Warguła, Ł., Krawiec, P., et al. (2018). Legal regulations of restrictions of air pollution made by non-road mobile machinery—The case study for Europe: A review. *Environmental Science and Pollution Research, 25*, 3243–3259. https://doi.org/10.1007/s11356-017-0847-8
117. Wang, Z., Huang, X., & Ding, A. (2018). Dome effect of black carbon and its key influencing factors: A one-dimensional modelling study. *Atmospheric Chemistry and Physics, 18*, 2821–2834. https://doi.org/10.5194/acp-18-2821-2018
118. WHO (World Health Organization). (2021). *WHO global air quality guidelines: particulate matter (PM2.5 and PM10), ozone, nitrogen dioxide, sulfur dioxide and carbon monoxide*. World Health Organization. https://apps.who.int/iris/handle/10665/345329. License: CC BY-NC-SA 3.0 IGO.
119. Wihersaari, H., Karjalainen, P., Pirjola, L., Saukko, E., Kuuluvainen, H., Kulmala, K., Keskinen, J., & Rönkkö, T. (2020). Particulate emissions of a modern diesel passenger car under laboratory and real-world transient driving conditions. *Environmental Pollution, 265*, art. no. 114948. https://doi.org/10.1016/j.envpol.2020.114948
120. Wilker, E. H., Preis, S. R., Beiser, A. S., Wolf, P. A., Au, R., Kloog, I., et al. (2015). Long-term exposure to fine particulate matter, residential proximity to major roads and measures of brain structure. *Stroke, 46*(5), 1161–1166. https://doi.org/10.1161/STROKEAHA.114.008348
121. Wong, J. P. S., Tsagkaraki, M., Tsiodra, I., Mihalopoulos, N., Violaki, K., Kanakidou, M., Sciare, J., Nenes, A., & Weber, R. J. (2019). Effects of atmospheric processing on the oxidative potential of biomass burning organic aerosols. *Environmental Science & Technology, 53*(12), 6747–6756. https://doi.org/10.1021/acs.est.9b01034
122. Wong, A. Y. H., Selin, N. E., Eastham, S. D., Mounaïm-Rousselle, C., Zhang, Y., & Allroggen, F. (2024). Climate and air quality impact of using ammonia as an alternative shipping fuel. *Environmental Research Letters, 19*(8), Article 084002. https://doi.org/10.1088/1748-9326/ad5d07
123. Xu, P., Liu, B., Chen, H., Wang, H., Guo, X., & Yuan, J. (2024). PAHs as environmental pollutants and their neurotoxic effects. *Comparative Biochemistry and Physiology Part C: Pharmacology, Toxicology and Endocrinology, 283*, 109975. https://doi.org/10.1016/j.cbpc.2024.109975
124. Yao, G., Yue, H., Yun, Y., & Sang, N. (2015). Chronic SO2 inhalation above environmental standard impairs neuronal behavior and represses glutamate receptor gene expression and

memory-related kinase activation via neuroinflammation in rats. *Environmental Research, 137*, 85–93. https://doi.org/10.1016/j.envres.2014.11.012
125. Zhang, Z., Wang, W., Cheng, M., Liu, S., Xu, J., He, Y., & Meng, F. (2017). The contribution of residential coal combustion to PM2. 5 pollution over China's Beijing-Tianjin-Hebei region in winter. *Atmospheric Environment, 159*, 147–161. https://doi.org/10.1016/j.atmosenv.2017.03.054
126. Zhu, Z. W., Yang, R. L., Dong, G. J., & Zhao, Z. Y. (2005). Study on the neurotoxic effects of low-level lead exposure in rats. *Journal of Zhejiang University. Science. B, 6*(7), 686–692. https://doi.org/10.1631/jzus.2005.B0686
127. Moreno-Ríos A. L., Tejeda-Benítez L. P., & Bustillo-Lecompte C. F. (2022). Sources characteristics toxicity and control of ultrafine particles: An overview, *Geoscience Frontiers,* 13(1). https://doi.org/10.1016/j.gsf.2021.101147

Chapter 3
How Do Air Pollutants Affect the Brain? Potential Entry Routes and Peripheral Effects

Laura Mussalo, Claire Fayad, Muhammad Ali Shahbaz, Elena Puris, and Katja Marika Kanninen

Keywords Air pollution · Ultrafine particles (UFPs) · Neuroinflammation · Particle translocation · Oxidative stress · Neurodegeneration

3.1 Introduction

Accumulating evidence demonstrates that air pollution exposure is linked to adverse effects on the brain, including associations with neurodevelopmental, neurodegenerative, and mental diseases. Particulate matter (PM), a main component of air pollution, is a complex mixture of solid particles and liquid droplets in the air we breathe. While most prior research has focused on the health effects of PM2.5 (particles with diameter less than 2.5 microns) and PM10 (particles with diameter less than 10 microns), results are also starting to emerge on the adverse brain effects caused by exposure to ultrafine particles (UFPs, with diameter less than 100 nanometers). These particles influence the biological barriers of the brain, including the blood-brain-barrier (BBB), the lung-brain barrier, and the olfactory-brain barrier, and induce neuroinflammation and oxidative stress in the brain. This chapter summarizes the brain cellular and molecular mechanisms induced by air pollutants, provides evidence for particle presence in the brain, and discusses the potential entry and clearance routes of particles in the brain.

L. Mussalo · C. Fayad · M. A. Shahbaz · E. Puris · K. M. Kanninen (✉)
A.I. Virtanen Institute for Molecular Sciences, University of Eastern Finland, Kuopio, Finland
e-mail: katja.kanninen@uef.fi

K. M. Kanninen, A. R. White (eds.), *Air Pollution and the Brain*,
https://doi.org/10.1007/978-3-031-99302-2_3

3.2 Air Pollutant Particles Adversely Affect the Brain via Distinct Cellular Mechanisms

3.2.1 Inflammation

Historically, the brain was regarded as an immune-privileged organ, protected from the body's immune surveillance by the blood-brain barrier (BBB) [1]. This view has since evolved, as new evidence reveals the critical roles that the immune system plays within the central nervous system (CNS). Immune cells actively maintain cerebral homeostasis, eliminating pathogens, and assisting in recovery from injuries. However, when inflammatory stimuli are not adequately cleared, resolution mechanisms can become compromised, leading to chronic inflammation and subsequent neurodegeneration [2]. Neuroinflammation involves a broad spectrum of immune responses within the CNS [3], which can be triggered by various factors, including endogenous protein aggregates, environmental toxins, and genetic mutations such as those in apolipoprotein E4 (APOE4) [4]. Persistent inflammation in the brain results in the sustained release of neurotoxic factors, and proinflammatory mediators, contributing to the pathology of several neurodegenerative diseases [5].

Recent studies have highlighted the impact of environmental factors, such as exposure to airborne pollutants, on neuroinflammation. These exposures induce several acute systemic responses, including increased oxidative stress [6], lymphocytosis, neutrophilia [7], and alterations in both innate and adaptive immunity [8]. Such inflammatory pathways can cause neuronal damage, in which two primary mechanisms are involved: direct infiltration of particles into the CNS, initiating a local inflammation, or indirectly through systemic inflammation [9]. The hallmark features of neuroinflammation include increased vascular permeability, leukocyte infiltration, glial cell activation, and the production of a wide array of inflammatory mediators, such as cytokines (IL-1β, IL-6, TNFα), chemokines (CCL2, CCL5, CXCL1), secondary messengers (nitric oxide and prostaglandins), and reactive oxygen species (ROS) [10].

Microglia, the intrinsic immune cells of the brain, possess multifaceted roles. They contribute to the integrity of the BBB, eliminate apoptotic cells, and perpetually monitor and respond to any cerebral changes to safeguard neurons and sustain tissue equilibrium [11]. Given the dynamic nature of microglia and their swift responsiveness to environmental cues, exposure to pollutants can modify their physiological characteristics and potentially contribute to pathological states. Thus, in a recent study, low concentration of PMs was linked to microglial morphological changes and activation, while the pollutants impaired microglial phagocytosis and induced the secretion of certain proinflammatory cytokines, marked by proinflammatory mRNA expression [11, 12].

Mechanistically, pollutant particles can activate microglia directly through a pattern recognition complex that includes microglial MAC1 and scavenger receptors [13]. Scavenger receptors facilitate the internalization of particles through endocytosis, triggering the activation of microglial NADPH oxidase NOX2 and subsequent production of ROS. Elevated microglial activity leads to the release of neurotoxic cytokines, particularly tumor necrosis factor-α (TNF- α), interleukin 1 (IL-1), and interleukin 6 (IL-6). These cytokines act as prominent inflammatory signaling mediators, fostering tissue swelling, the release of cytotoxic compounds by immune cells, and further amplification of pro-inflammatory signals, contributing to widespread neuroinflammation in the brain, leading to damage and diffuse neural tissue loss across various brain regions [14]. This includes the impairment of dopaminergic neurons function, which is physiologically critical for cognitive functions [14]. Supporting this, exposure of rats to nanoparticle substances was found to trigger dementia-like clinical symptoms, accompanied by heightened levels of neuroinflammatory mediators such as Aβ-42, ROS, hydrogen peroxide, and NO_2. Similarly, a recent investigation revealed that exposure of mice to nanoparticle substances derived from traffic led to increased neuronal amyloidogenesis through oxidative damage, indicating a close association with dementia [15].

Another crucial actor in the brain is the astrocyte, which participates in the formation and maintenance of the BBB, controlling the transport of molecules, providing nutrients and physical support for neurons, and removing excess neurotransmitters from the synaptic cleft [16]. The interplay between microglia and astrocytes involves a complex and intimate crosstalk, where these cell types influence each other's activity through the release of various molecules such as interleukin 1α (IL-1α), TNFα, and C1q. Microglia can influence astrocytic activation states, directing them towards a more proinflammatory profile, while astrocytes can also modulate the function of microglia. Hence, exposure to traffic-related air pollution can lead to astrocytic activation and reactive astrogliosis, indicated by increased glial fibrillary acidic protein (GFAP) expression levels in astrocytes [17]. This activation can result in astrocyte swelling, changes in molecule transportation, and ion homeostasis, which may contribute to brain dysfunction. The crosstalk between microglia and astrocytes will induce activation of both cell types, leading to increased expression of proinflammatory cytokines and altered glial cell levels [17]. In an experimental study by Yan et al., the inhalation of NO_2 resulted in a modulation of prostaglandin E2 via cyclooxygenase 2, impacting the function of both astrocytes and microglia, inducing a process akin to Alzheimer's disease [15]. Thus, the interplay between microglia and astrocytes in response to air pollution can synergistically contribute to chronic inflammation and neurodegeneration in the brain.

3.2.2 *Oxidative Stress*

Oxidative stress is a term describing an imbalance between oxidants and antioxidants, a consequence of unbalanced cellular redox homeostasis, which manifests as damage to biomolecules, leading eventually to permanent disruption in cellular functions [18]. One of the most prominent effects in the brain caused by air pollution is oxidative stress [19, 20]. Dysfunctional mitochondria are one core reason for oxidative stress. Not only are they especially vulnerable to environmental toxicants, but their dysfunction is also shown to have a major role in the development and progression of neurological diseases [21–23]. Given the high energy demand of the brain, neuronal tissue is greatly dependent on ATP production via oxidative phosphorylation; therefore, carefully orchestrated mitochondrial functions and redox balance are essential [24, 25].

Exposure to air pollutants has been linked with increased oxidative stress in several cell models [26–29]. For example, we have earlier shown that exposure to PM2.5 in vitro causes elevation of reactive oxygen species (ROS) in human olfactory mucosa cells, located in the nasal cavity [26]. In addition to elevated ROS levels, also perturbed antioxidant defense mechanisms have been demonstrated, including alterations in glutathione metabolism and decreased superoxide dismutase (SOD) activities [27–29]. When assessing cell models to investigate oxidative stress, it should be remembered that cells *in vitro* generate more ROS than cells *in vivo,* mainly due to the lack of antioxidants and high concentration of oxygen in cell culture conditions [30]. However, cell models provide important insight into molecular-level mechanisms of how different air pollutants induce oxidative stress. These models also enable a more specific understanding of the complexity of redox chemistry behind oxidative stress [18].

Several studies have assessed the impact of air pollution exposure on oxidative stress in the brain, some of which have also carried out behavioral studies to decipher alterations to behavioral and cognitive functions [31–33]. Interestingly, exposure to UFPs has been shown to perturb redox homeostasis via the methionine–glutathione pathway and cause oxidative stress in the hippocampus [34, 35]. Furthermore, rats exposed to simulated vehicle emission exhausts manifested neurological deficits accompanied by increased oxidative stress, diminished antioxidant response, and mitochondrial impairment [32].

In epidemiological studies, air pollution-mediated oxidative stress has been suggested to affect childhood developmental outcomes, ranging from changes in brain structures to subclinical deficits and even to developmental disorders [36], such as autism [37]. PM has been shown to induce oxidative stress by reducing the function of the antioxidant system in schizophrenia patients, leading to increased relapse risk and hospital admission [38]. Post-mortem studies have revealed a connection between air pollution and oxidative stress coupled with Alzheimer's and Parkinson's-like changes in the brains of children and young adults living in a polluted city [39, 40]. Taken together, it is evident that oxidative stress is one of the core reasons for air pollution-mediated adverse effects on the brain. Thus, more research is undoubtedly needed for a better understanding of the molecular mechanisms behind these effects to develop mitigation therapies.

3.3 Evidence for Particle Presence in the Brain

Particle presence in the human brain was reported for the first time in 1992 by Kirschvink et al. [41]. This study examined post-mortem brains and revealed a wide distribution of ferromagnetic material in various human brain tissues covering cerebral lobes, the cerebellum, the basal ganglia, and the midbrain. Kirschvink and colleagues concluded that these particles are most likely to be formed within human tissues rather than from an environmental origin. Similar findings were done by others, verifying that magnetite in the brain has a connection to changes related to Alzheimer's disease [42, 43]. However, later it was discovered that these particular magnetite particles residing in the brain have to originate from an external source instead of forming in the brain [44]. This finding led to a whole new direction in investigating particle accumulation in the brain. Several animal models have shown the building of particles in the brain upon exposure to air pollutants [45, 46], and especially upon exposure to the smallest nano/ultra scale particles [38]. A 3D human brain model showed an accumulation of PM2.5 in the brain tissue [47], and recently the occurrence of black carbon in the post-mortem brain of AD patients has been demonstrated [48]. In addition, children and young people living in highly polluted areas exhibit signs of brain inflammation, volumetric and metabolic brain changes, attention and short-term memory deficits, and hallmarks of AD and Parkinson's disease (PD) [49].

3.4 Deposition of Pollutant Particles at Different Locations of the Respiratory Tract

PMs, characterized by their size and aerodynamic properties, are more than just a collection of microscopic particles suspended in the air; they serve as carriers for a myriad of chemicals, creating a complex mixture that infiltrates our respiratory systems with every breath inhaled [50]. The respiratory system, comprising the upper and lower airway passages, includes structures from the nose and mouth down to the alveoli. By the lung geometry, particulates deposit through interception, impaction, sedimentation, or diffusion mechanisms, influenced by factors like size, shape, and density [51]. Thus, understanding the intricate relationship between PM and systemic effects through the distribution of particles in bodily systems is imperative for mitigating the risks posed by air pollution.

PMs come in various sizes, each with its own unique ability to navigate pathways in the respiratory tract. From total suspended particles to UFPs, PMs can be classified based on their aerodynamic equivalent diameter (Dp): total suspended particle (Dp < 100 μm), inhalable particle (PM10, Dp < 10 μm), fine particulate matter (PM2.5, Dp < 2.5 μm), and ultrafine particulate matter (UFP, Dp < 0.1 μm), with each size category posing distinct risks to respiratory health [52]. While larger particles (>10 μm) are typically trapped in the nose and throat due to impaction and usually cleared by mucociliary mechanisms, inhalable particles, such as PM10, primarily deposit in the upper airways, including the nasopharynx and larynx [53]. Sedimentation occurs when gravitational forces overcome particle buoyancy, causing particles between 2.5 and 10 μm, known as the coarse fraction, to settle in the tracheobronchial tree [50].

However, nanoparticles, ranging from 1 to 100 nanometres, have the remarkable ability to penetrate deep into the alveoli, where gas exchange occurs [54]. These particles, including UFPs with diameters less than 0.1 μm, can evade the body's natural filtration mechanisms, paving their way into the deepest recesses of the lungs by diffusion [14]. Due to their small size and the complex airflow patterns within the nasal passages, UFPs can also deposit in the nasal cavity [55, 56]. This deposition is significant because the nasal cavity serves as a primary entry point to the respiratory tract and contains a rich network of olfactory receptors. This anatomical arrangement potentially allows for the translocation of particles to the brain via the olfactory mucosa located at the roof of the nasal cavity [57, 58].

UFPs can also deposit in the airways and centriacinar regions of the lung, triggering airway diseases due to their difficult clearance by alveolar macrophages. This delayed clearance allows UFP to enter the pulmonary interstitial sites, inducing pulmonary inflammation and potentially entering the systemic circulation upon absorption into the bloodstream [59]. For instance, studies using labeled carbon UFPs detected radioactivity in the blood shortly after inhalation, confirming their systemic absorption [60]. Despite their small size, the extensive surface area of UFPs facilitates the absorption of toxic compounds, including heavy metals and organic chemicals, thereby increasing their toxicity and potential harm to the respiratory system. Consequently, the diameter of particulate matter inversely correlates with its potential to induce adverse systemic and neurological effects [61].

3.5 Particle Effects on Biological Brain Barriers

3.5.1 Olfactory-Brain Barrier

The epithelial tissues of the body act as a regulatory barrier covering all bodily surfaces that are exposed to the external environment. All substances that enter the body must cross this strictly restricted barrier, which regulates secretion, absorption, and transport across surfaces. The location of the epithelium determines its specific structure and adaptation to specific functions. Structures contributing to epithelial barrier function include cell-cell junctions (tight junctions [TJ], adherens junctions [AJ], and desmosomes), actin filament organization, stiffness of extracellular matrix (ECM), and organization of integrins [62–64].

Upon exposure to pollutants, several studies have been conducted *in vitro* with the human nasal epithelial cell (HNEC) RPMI 2650 cell model [65, 66] and with human-derived nasal epithelial cells [67–71], and with animal models [66, 67]. These studies have shown compromised epithelial barrier function through decreased expression of TJ proteins, such as zona occludens-1 (ZO-1), occludin, claudin-1, and e-cadherin, reduced trans-epithelial electrical resistance (TEER), increased endoplasmic reticulum stress, oxidative stress, and release of proinflammatory cytokines. However, the effects of air pollutants on human olfactory mucosa and its integrity have been far less studied. We have previously assessed the effects of ultrafine particles on a human-derived in vitro model of OM, where RNA-Seq and functional level examination revealed that UFPs impair barrier function [72]. Taken together, these data suggest a compromise of the olfactory-brain barrier upon exposure to pollutants.

In mice, exposure to diesel exhaust particles [73] and to urban particulate matter (UPM) [74] has been shown to disrupt olfactory signaling networks, indicating that UPM is toxic to the nasal and olfactory epithelia, which are barriers covering the olfactory neuroepithelium. Local inflammation in the nasal cavity can also elicit behavioral abnormalities in mice [75]. However, more research is needed to determine the impact of air pollution exposure on the olfactory-brain barrier *in vivo*.

3.5.2 Lung-Blood Barrier

Once the PM is infiltrated into the different regions of the lungs, deposits lingering on the endothelial cells can impair the functions of the lung barrier, further amplifying the particle translocation in a self-inducing manner [76]. Effects of traffic-related emissions on the human respiratory system have been investigated widely *in vitro*. Most of the studies have been done with lung cell models, with human bronchial epithelial cell lines. The main mechanisms of how air pollutants disrupt the barrier function in these models include decreased expression of epithelial TJ proteins: ZO1, Occludin, and E-cadherin, inflammatory responses, and downregulation of filaggrin [77], oxidative stress [78], ciliary dysfunction, hyperplasia of goblet cells, inflammation, and even cytotoxicity with the highest doses [79]. Disruption of epithelial barrier function in lungs may also lead to altered immune responses and increased susceptibility to infection [80].

3.5.3 Blood-Brain-Barrier

The normal functioning of the CNS, with its highly controlled microenvironment, is maintained by the highly protective blood-brain barrier (BBB) separating the brain from systemic circulation. The BBB is formed by the cerebral microvascular endothelial cells, which cooperate with the other cells of the neurovascular unit (NVU), such as astrocytes, pericytes embedded within the capillary basement membrane, microglia, and neurons [81]. The BBB is a dynamic barrier with several specific features allowing it to defend the brain parenchyma from blood-borne agents. These characteristics include the presence of tight junctions (TJ) between the cerebral microvascular endothelial cells, lack of fenestrations, expression of metabolizing enzymes and membrane transporters responsible for regulating the passage of solutes across the BBB [81]. The transmembrane TJ proteins consist of three integral membrane proteins, such as claudin, occludin, and junction adhesion molecules, as well as several cytoplasmic accessory proteins, including Zonula occludens-1, -2, -3 (ZO-1, ZO-2, ZO-3) [82]. The TJs seal the paracellular route between adjacent microvascular endothelial cells and thereby restrict the passage of water-soluble molecules across the BBB. In addition, the expression of metabolizing enzymes in the brain microvascular endothelial cells, including those from the cytochrome P450 superfamily, represents a metabolic barrier responsible for eliminating toxic compounds and metabolites from the brain [83]. Finally, membrane transporters of the ATP-binding cassette (ABC) and solute carrier (SLC) superfamilies, which are highly expressed in the brain microvascular endothelial cells, play a key role in regulating the uptake of essential nutrients and efflux of unwanted compounds [84, 85]. A disturbance of the function of BBB components can lead to BBB disruption with a subsequent cascade of neuroinflammatory events resulting in neuronal dysfunction and degeneration [86].

Under physiological conditions, the intact BBB is supposed to protect the brain from the entry of exogenous toxicants, including air pollutants. However, studies on engineered NPs and advances in research on drug delivery to the brain via NPs suggest that inhaled combustion-generated UFPs and exogeneous NPs can cross the intact BBB via adsorptive- and receptor-mediated transcytosis as well as carrier-mediated transport [87]. The smallest airborne PM can enter the systemic circulation from the alveoli, where they may cross the BBB. Indeed, Lu et al. (2020) detected the inhaled exogeneous NPs in human serum [88]. Moreover, Qi et al. (2022) found exogeneous fine particles in human cerebrospinal fluid (CSF) and blood, providing evidence of potential entry of inhaled fine particles from the systemic circulation to the brain [60]. In rats, in addition to direct entry through the olfactory nerves, the inhaled NPs reached the brain via the lung-blood-BBB path, although the nasal NP delivery route had a significantly greater impact on NP accumulation in the brain [89].

In addition, the entry of PM can occur when the BBB function is impaired in a disease state or due to direct exposure to PM, circulating in the blood, and interacting with the BBB at its luminal site. There is growing evidence that circulating PM can modulate the structural integrity of the BBB and increase its permeability, leading to brain dyshomeostasis and worsening of CNS disease progression. For instance, a mechanistic study by Qi et al. (2022) demonstrated BBB permeation of exogenous particles from the blood after crossing the air–blood barrier with consequent damage to the BBB, such as a decrease of the ZO-1 level after particle treatment in vitro [60]. In addition, the findings from epidemiological studies highlight the potential effect of air pollution on BBB integrity. Thus, long-term exposure to air pollution in Mexico City subjects led to decreased expression of ZO-1 in the frontal cortex [39]. In another study, the children highly exposed to air pollution had greater levels of serum high-affinity IgG and IgA actin and occludin-zonulin antibodies compared to those in the low air pollution exposure group, indicating BBB breakdown [90].

The effect of traffic-generated air pollutants on the brain microvasculature was investigated in animal models. In male atherosclerotic Apolipoprotein (Apo) $E^{-/-}$ and C57BL/6 mice, the exposure to traffic-generated air pollution increased BBB permeability and caused BBB disruption as shown by altered expression of microvascular occludin and claudin-5, associated with a concomitant increase in activity of matrix metalloproteinases 2 and 9 [91, 92]. Later, Suwannasual et al. (2019) revealed that changes in BBB integrity due to inhalation exposure to traffic-generated pollutants in C57BL/6 mice were mediated by Angiotensin (Ang) II signalling through the Ang II type 1 receptor and inflammation, which were worsened by a high-fat diet [93]. Similarly to these studies, inhalation exposure to mixed vehicle emissions affected BBB integrity and permeability in female atherosclerotic ApoE$^{-/-}$ mice independently of ovary/hormone status [94]. Thus, exposure to mixed vehicle emissions in these mice resulted in reduced expression of occludin and claudin-5, increased IgG leakage into the brain parenchyma, and upregulated expression of endothelial cell adhesion molecules, indicating activation of endothelial cells [94]. In addition, prenatal and early life exposure of rats to traffic-generated air pollutants represented by nanoparticulate matter (nPM; aerodynamic diameters $\leq$ 200 nm) from gestation into maturity led to hippocampal subregional alterations in ZO-1 expression, which was associated with microbleeds [95]. Kang et al. (2021) demonstrated that the treatment of human in vitro BBB model, i.e. hCMEC/D3 cells, with fine PM2.5 (aerodynamic diameters $\leq$ 2.5 μm) significantly reduced the expression of ZO-1 [47]. Furthermore, air pollution can worsen brain vascular dysfunction resulting from pathological conditions. For example, the exposure to nPM from urban traffic exacerbated the BBB permeability in a murine bilateral carotid artery stenosis mouse model of chronic cerebral hypoperfusion, as demonstrated by increased extravascular IgG deposits [96].

In summary, although there is evidence of BBB-mediated entry of air pollutants into the brain and their effect on BBB functional and structural integrity, more mechanistic studies are required to reveal the transport pathways of inhaled PM from the blood to the brain across the BBB as well as the role of BBB damage in this process. In addition, most of the studies have investigated the impact of air pollution on TJs and changes in paracellular permeability, while future studies should focus on the potential effect of exposure to air pollutants on other components of the BBB, including metabolizing enzymes, receptors, and transporters.

3.6 Proposed Mechanisms of Particle Brain Entry

3.6.1 Systemic Circulation

Particles can enter the systemic circulation mainly through the lower respiratory tract, comprising the trachea, primary bronchi, and lungs (including the alveoli), and through the gastrointestinal tract [97]. From the lungs, particles can either directly bypass the lung-blood barrier or indirectly induce systemic inflammatory or immune responses. Some of the particles may also end up in the lymph as phagocytosed by cells such as the pulmonary macrophages [76, 97]. A study done with human volunteers and mice showed that inhalation of gold nanoparticles in the lung led to elevated concentrations of gold particles in blood and liver, demonstrating that nanoparticles can translocate from the lung to the circulation [98]. In addition, a mouse study with exposure to particles from burning candles promoted telomere shortening in the lungs and spleen, and accelerated the progression of atherosclerosis in the aorta, further implying that particles can reach other organs via lung exposure [99]. A study done on human volunteers demonstrated that orally given nanosized titanium dioxide translocated to the bloodstream through the gut [100]. PM has also been shown to increase inflammatory reaction and gut permeability *in vitro* and *in vivo* models [101–103], implicating the gut to be a possible route for PM to enter the systemic circulation. After intruding into the systemic circulation, particles may translocate even to the brain by penetrating the BBB [104]. Qi and colleagues demonstrated in mice the translocation of intratracheally administered PM to various regions of the brain, coupled with particle aggregates near the damaged BBB structures [60].

3.6.2 Olfactory Nerve

The olfactory mucosa (OM), residing in the rooftop of the nasal cavity, is an innervated tissue which is directly exposed to the ambient air we breath [105]. As a sensory tissue, olfactory sensation occurs in this system, where the chemical stimulus is translated into signals and conducted through the olfactory nerves (I cranial nerve) to the olfactory bulb (OB) and further to the brain [106, 107].

Oberdörster et al. were the first to demonstrate ultrafine particles translocating through the olfactory nerve to the central nervous system in rats exposed to ultrafine elemental 13C particles [108]. This was supported by the findings of Elder and colleagues, who concluded that the olfactory neuronal pathway can efficiently translocate inhaled Mn oxide as solid UFPs to the central nervous system, and despite the differences between species, this pathway is also relevant in humans [109]. Rodents are obligatory nose breathers [105], therefore, the execution of rodent inhalation studies with whole-body inhalation chambers gives an almost explicit view of exposure through the olfactory system, however, some other mechanisms for particle entry may occur. It is shown in mice that the oxidative stress and inflammatory responses caused by exposure to pollutants are mediated from the olfactory epithelium into the brain [110]. In addition, it has been visually demonstrated in mice that inhaled ultra-fine soot particles reached the brain via olfactory nerves and OB, accompanied by neural inflammation [111]. Moreover, exposure to PM2.5 has been shown to trigger synaptic dysfunction in the olfactory brain route in mice, suggesting it is the major pathological feature leading to cognitive impairment [112].

Findings from Mexico City children and young adults have shown that high levels of air pollution accumulate in the olfactory epithelium and OM, and translocate through the olfactory nerve to the OB [113, 114]. Computational simulations have shown that inhaled nanoparticles of size 1–2 nm deposit most efficiently in the olfactory region in humans, and when the particle size increases up to 100 nm, the deposition decreases [56]. However, Maher and colleagues demonstrated that magnetite particles with a diameter < 200 nm can access the brain directly through the olfactory nerve [44], and very recently, black carbon particle presence in the brain structures, such as prefrontal cortex, including OB, thalamus, and hippocampus, was shown in post-mortem AD-patients [48]. This evidence clearly highlights the importance of the olfactory route as a portal for particles into the brain.

3.6.3 *Trigeminal Nerve*

The trigeminal nerve, the fifth paired cranial nerve, plays a crucial role in nociception through its specialized nerve endings in the nasal mucosa [97]. It innervates the nasal cavity and projects to the spinal trigeminal nucleus, serving as a pathway for the transport of inhaled aerosol particles directly from the nasal epithelium to the brain, potentially leading to neurotoxic effects such as inflammation and cellular damage [104, 115]. Inhaled particles can deposit on the olfactory epithelium and respiratory mucosa, interacting with the sensory nerve endings of the trigeminal nerve located in the nasal epithelium, and enabling their translocation to the CNS [56]. The ophthalmic and maxillary branches of the trigeminal nerve, which primarily innervate the nasal and sinus regions, can retrogradely transport particles from the periphery towards the CNS [116]. Trigeminal nerve fibers project into the brainstem, specifically to the trigeminal ganglion and sensory nuclei, where pollutant particles may initiate neurotoxic effects [39, 117].

Animal studies have demonstrated that particles instilled intranasally can reach the brain, particularly in regions connected by the trigeminal nerve, within hours of exposure. Lewis et al. showed elevated manganese levels observed in the trigeminal ganglia and spinal trigeminal nucleus of rodents shortly after inhalation [118]. Similar findings were reported by Oberdorster et al. and Gianutsos et al., who observed increased concentration of UFPs in the trigeminal ganglia following controlled exposure [108, 119]. The direct synaptic innervation of the nasal epithelium by the trigeminal nerve suggests a direct route for toxicants to enter the CNS [104]. Epidemiological studies indicate that populations in areas with high air pollution have higher rates of neurological disorders, supporting the hypothesis that pollutants may reach the CNS via the trigeminal nerve [39]. However, extensive epidemiological research on this specific pathway remains limited, underscoring the need for further studies to better understand the underlying mechanisms.

3.6.4 *Neuroenteric Pathway*

The intricate relationship between the gastrointestinal (GI) tract and the central nervous system (CNS) has garnered significant attention in recent years. The bidirectional communication pathway, known as the gut-brain axis (GBA), plays a crucial role in maintaining homeostasis of gut functions and influencing various physiological processes, including emotional and cognitive function [120]. In recent years, with the gradual deepening of research, researchers have expanded their horizons beyond digestive tract diseases and found an inseparable link between the gut microbiota and the brain in neuropsychiatric ailments like depression and autism spectrum disorders, as well as neurodegenerative diseases like Parkinson's disease (PD), Alzheimer's disease (AD), and amyotrophic lateral sclerosis (ALS) [121–124]. For example, chronic GI inflammation and dysbiosis have been associated with an increased risk of AD and PD. In AD, altered gut microbiota and the resultant inflammation may promote amyloid-beta deposition and neuroinflammation in the brain [125]. Similarly, in PD, GI tract inflammation is thought to trigger alpha-synuclein aggregation, which can then spread to the brain via the vagus nerve [126].

A recent shift in the focus of pollution research from the respiratory tract to the GI tract has revealed evidence suggesting that the GBA may serve as a gateway for the translocation of PM and environmental toxins from the GI tract to the brain, thereby potentially impacting neurological health [60, 127]. Sensory neurons within the enteric nervous system detect the presence of PM particles in the GI tract or respond to alterations in the GBA, activating afferent vagal fibers that transmit these signals to the brain. This pathway provides a potential route for ingested pollutants to directly influence CNS function. The PM can enter the GI tract through several mechanisms. Primarily, PM is inhaled into the respiratory system, where a portion is swallowed and subsequently passes into the GI tract. Additionally, PM can settle on food and drinks, leading to ingestion. It is estimated that approximately 5% of inhaled PM is ingested and enters the GI tract. This ingestion contributes to the systemic distribution of PM, where it can exert localized effects on the gut and potentially translocate to other organs, including the brain.

Once ingested, the PM and other environmental toxins can significantly disrupt the GBA. When PM and toxins interact with the gut microbiota—the diverse community of microorganisms in the GI tract—dysbiosis can ensue, and this imbalance in the microbial population can trigger systemic inflammation and oxidative stress. Such inflammatory responses can compromise the gut barrier integrity, facilitating the entry of toxins into the bloodstream and thereby potentially reaching the brain. For example, exposure to PM2.5 has been shown to alter the gut microbiota composition, reducing beneficial bacteria like Faecalibacterium while increasing harmful bacteria such as Escherichia/Shigella and Akkermansia. These microbial changes are associated with cognitive impairments and elevated systemic inflammation, indicating that the GBA plays a crucial role in mediating these adverse effects [128].

The neuroenteric pathway involves the translocation of particles from the GI tract to the CNS, with the vagus nerve serving as a critical conduit. Sensory neurons in the enteric nervous system detect the presence of PM particles, activating afferent vagal fibers that transmit these signals to the brain. This pathway provides a route for ingested pollutants to affect the CNS directly. Animal studies support this concept; nanoparticles administered orally have been detected in the brain, suggesting their possible translocation via the neuroenteric pathway. Furthermore, a recent study suggests that exposure to combustion-derived nanoparticles in Mexico City may contribute to the development of early PD by damaging the GI barrier and affecting the nervous system. The combustion-derived nanoparticles were found abundantly in various cells in the GI tract, parasympathetic fibers, and the vagus nerve, and associated with pathological changes similar to those seen in PD and AD [49]. Moreover, exposure to PM2.5 in the GI tract induces local inflammation and oxidative stress, leading to the release of inflammatory mediators in the blood and altering the gut microbiome [129, 130] via alterations in the GBA. These mediators can activate the vagus nerve, triggering neuroinflammatory responses in the brain.

In conclusion, the neuroenteric pathway represents a fascinating yet complex route for potential particle translocation to the brain. While some research has been focused on understanding this interaction, the mechanism behind it remains poorly understood.

3.6.5 Placental Pathway

The placenta serves as a critical interface between the maternal and fetal environments, facilitating nutrient exchange, waste removal, and protection against harmful substances. Emerging evidence suggests that air pollution particles, including PM, can translocate across the placental barrier, potentially reaching the fetal compartment.

Animal studies have provided insights into how nanoparticles such as gold, silica, and titanium dioxide can cross the placental barrier and potentially impair fetal development and cause neurotoxicity [131–136]. However, the placenta is a highly species-specific organ, and results from rodent models do not directly translate to humans.

Recent studies in humans have suggested that ambient fine PMs can cross the human placental barrier from the maternal circulation [137, 138]. The presence of PMs on the fetal side of the placenta indicates their potential to circulate in fetal blood and directly affect all fetal organ systems [138]. These particles can bypass the placental barrier via passive diffusion, primarily influenced by factors such as particle size, shape, and surface charge [137]. Ex vivo human placental perfusion models offer an ethically viable and accurate representation of in vivo conditions for studying the placental transport of xenobiotics and nanoparticles. These models have also demonstrated the presence of particles in placental tissues, supporting the feasibility of passive diffusion as a mechanism of particle transport [139–141].

In addition to the direct effect of the translocated particles on the fetus through placental transport, inhaled air pollutants deposited in the mother's lungs can induce oxidative stress and inflammatory responses in the placenta, affecting its function [142, 143]. This inflammatory response can impair the placental ability to transport oxygen and nutrients to the fetus, leading to underdevelopment of fetal brain and other organs [144]. Furthermore, inflammatory cytokines produced in response to air pollutants in the maternal circulation may be transported to the fetal blood, causing systemic inflammatory responses that delay fetal development [142, 143].

In summary, the translocation of air pollution particles to the fetal brain via the placental pathway raises concerns about potential neurodevelopmental effects of exposure. Studies suggest that prenatal exposure to air pollutants, including PM and nanoparticles, may contribute to neuroinflammation, oxidative stress, and alterations in brain structure and function. These effects may manifest as cognitive impairments, behavioral disorders, and increased susceptibility to neurodevelopmental disorders later in life. Therefore, understanding the placental pathways of air pollution is crucial for assessing the risks and developing strategies to mitigate the impact on fetal development and health.

3.7 Clearance of Pollutant Particles From the Brain

The ability of the brain to clear PM is crucial for maintaining overall neurological health and functioning. The glymphatic system, a clearance system unique to the brain, plays a vital role in removing various waste products, including PM [145]. However, inefficient clearance of particles can lead to the accumulation of harmful substances, contributing to persistent neuroinflammation and neuronal damage, which are precursors to neurodegenerative diseases. Thus, understanding how PM clearance pathways operate in the brain is essential for comprehending their long-term effects on neurological health.

Recent research has shed light on the brain's specialized clearance system, which functions independently of traditional lymphoid tissue. This system relies on a complex interplay of structures and processes. The glymphatic system, along with meningeal lymphatic vessels, facilitates the flow of cerebrospinal fluid (CSF), thereby facilitating the removal of waste from the interstitial spaces between brain cells [146]. This process is particularly active during sleep when the interstitial space expands, enhancing waste removal [147]. Cerebral arterial pulsation and smooth muscle function propel CSF circulation, ensuring efficient removal of both small molecules and biomacromolecules from brain tissue. Along with the BBB [148] and the involvement of astrocytes [149] and microglia [150], these elements maintain a stable biochemical environment within the brain.

Disruption of this delicate balance can lead to pathological changes within the brain. As discussed earlier, PM can affect the BBB, microglia, and astrocyte activity, resulting in neuronal damage and disrupting the brain's natural clearance processes. Although the literature does not yet provide a clear link, it can be inferred that PM will have a longer retention period in the brain, where they can accumulate and exacerbate their harmful effects. For instance, black carbon particles exhibit a longer half-life in the brain compared to other organs, potentially leading to neurotoxic effects, shedding light on the clearance impairment of particles from the CNS [60].

Depending on their size, UFPs after gaining access to the brain may either remain in the brain parenchyma or drain through pathways typically used for protein waste elimination [151]. It remains unclear whether these particles exit the brain via CNS waste disposal systems or accumulate within the parenchyma by impairing clearance pathways. Yet, studies utilizing isotope labeling have demonstrated the transport of inhaled fine particles from the lungs to the brain via blood circulation, highlighting a slower clearance rate in the brain compared to other organs [60]. From this perspective, enhancing brain fluid drainage pathways may offer therapeutic benefits in mitigating the adverse effects of UFPs on brain health. Thus, scientists are actively exploring strategies to promote brain health and potentially prevent or slow the progression of neurodegenerative diseases by optimizing sleep, maintaining a healthy lifestyle, and developing targeted therapies to support the clearance of pollutant particules from the brain and preserve cognitive functions.

Figure 3.1 summarizes the means by which pollutant particles can enter the human body and their subsequent biological impacts.

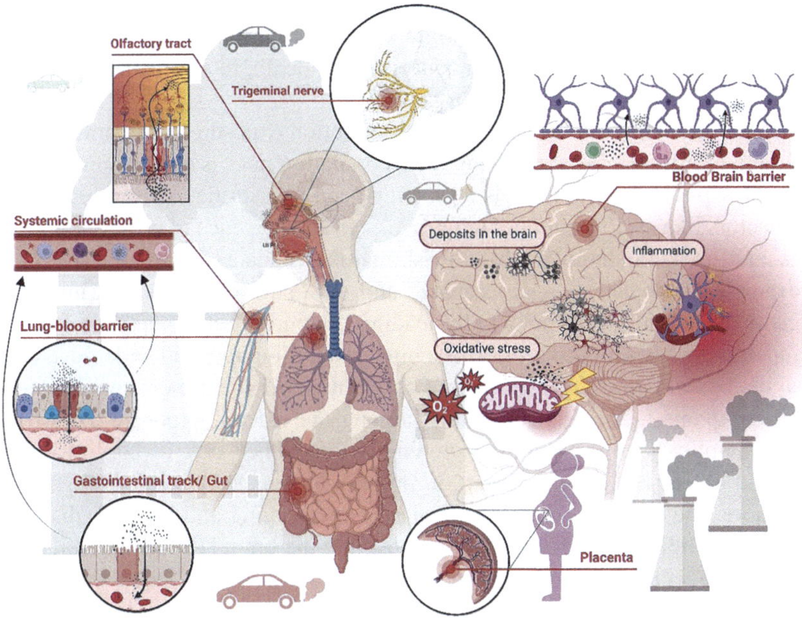

Fig. 3.1 Schematic representation of the pathways through which pollutant particles can enter the human body and their subsequent biological impacts. UFPs, primarily derived from sources such as vehicle emissions and industrial pollution, are inhaled and can cross the lung-blood barrier, entering systemic circulation. They may also access the central nervous system via the olfactory tract and trigeminal nerve, bypassing the blood-brain barrier. Additionally, UFPs can be absorbed through the gastrointestinal tract. Once in circulation, these particles accumulate in the brain, leading to oxidative stress, neuroinflammation, and neuronal damage. UFP exposure is also shown to affect the placenta, highlighting potential risks to fetal development during pregnancy. (Created in BioRender. Mussalo [158])

3.8 Conclusions and Future Research Needs

Significant strides have been made in the past few years to assess the effects of air pollution exposure on the brain and its links to brain disorders. However, there are several future research needs in this area. In particular, further research is needed to determine which components of air pollution are most harmful to the brain, and exactly how they confer their effects. It will be important to produce more evidence for pollution-disease associations and the underlying causality mechanisms and biological pathways. For instance, deciphering the entry and clearance mechanisms of various pollutants to the brain will be an important step forward. Once the mechanisms and biological pathways are better understood, it may be possible to design and test strategies to limit the adverse effects to confer brain protection against air pollution. Early identification of exposure health effects with the use of measurable biomarkers will help to predict future adverse health outcomes, enabling preventive actions and the use of mitigation strategies. Furthermore, adopting a life-course perspective and determining the impact of combined environmental exposures to e.g. air pollution and noise, can shed light on the health impacts of the totality of exposures a person experiences from conception to death, in a so-called exposome approach.

Susceptibility to the health impacts of air pollution exposure varies among individuals and groups. While considering total-population risks is a valid approach for safeguarding public health, there is a growing need to prioritize vulnerable groups [152]. Factors like socioeconomic status, age, sex, and presence of pre-existing conditions all affect an individual's health risk that is associated with exposure [153]. For example, considering age, prenatal and postnatal exposure detrimentally impacts neurodevelopment, potentially leading to neurodevelopmental and cognitive impairments [154]. Furthermore, the higher breathing rate and underdeveloped natural biological barriers in the lungs of children render them more susceptible to inhaled pollutants [155]. Concomitantly, among the elderly, exposure to PM has profound implications for health, including a heightened risk of cardiovascular events and increased accumulation of protein aggregates in the brain [156]. Furthermore, the BBB is known to be altered upon aging and in brain diseases, potentially facilitating the passage of particles into the brain [157]. Therefore, protection of vulnerable groups against the health impacts of air pollution should be a future priority. In the long term, key research findings could be implemented into future regulatory requirements for emission characterization and eventually lead to new standards and limit values for air pollution.

References

1. Fenster, R. J., & Eisen, J. L. (2017). Checking the brain's immune privilege: Evolving theories of brain–immune interactions. *Biological Psychiatry, 81*(2), e7–e9. https://doi.org/10.1016/j.biopsych.2016.10.027

2. Zhang, W., Xiao, D., Mao, Q., & Xia, H. (2023). Role of neuroinflammation in neurodegeneration development. *Signal Transduction and Targeted Therapy, 8*(1), 267. https://doi.org/10.1038/s41392-023-01486-5
3. Lyman, M., Lloyd, D. G., Ji, X., Vizcaychipi, M. P., & Ma, D. (2014). Neuroinflammation: The role and consequences. *Neuroscience Research, 79*, 1–12. https://doi.org/10.1016/j.neures.2013.10.004
4. Peggion, C., Calì, T., & Brini, M. (2024). Mitochondria dysfunction and neuroinflammation in neurodegeneration: Who comes first? *Antioxidants., 13*(2), 240. https://doi.org/10.3390/antiox13020240
5. Bevan-Jones, W. R., Cope, T. E., Jones, P. S., et al. (2020). Neuroinflammation and protein aggregation co-localize across the frontotemporal dementia spectrum. *Brain, 143*(3), 1010–1026. https://doi.org/10.1093/brain/awaa033
6. Mudway, I. S., Kelly, F. J., & Holgate, S. T. (2020). Oxidative stress in air pollution research. *Free Radical Biology & Medicine, 151*, 2–6. https://doi.org/10.1016/j.freeradbiomed.2020.04.031
7. Behndig, A. F., Larsson, N., Brown, J. L., et al. (2011). Proinflammatory doses of diesel exhaust in healthy subjects fail to elicit equivalent or augmented airway inflammation in subjects with asthma. *Thorax, 66*(1), 12–19. https://doi.org/10.1136/thx.2010.140053
8. Glencross, D. A., Ho, T. R., Camiña, N., Hawrylowicz, C. M., & Pfeffer, P. E. (2020). Air pollution and its effects on the immune system. *Free Radical Biology & Medicine, 151*, 56–68. https://doi.org/10.1016/j.freeradbiomed.2020.01.179
9. Block, M. L., & Calderón-Garcidueñas, L. (2009). Air pollution: Mechanisms of neuroinflammation and CNS disease. *Trends in Neurosciences, 32*(9), 506–516. https://doi.org/10.1016/j.tins.2009.05.009
10. Ji, R. R., Xu, Z. Z., & Gao, Y. J. (2014). Emerging targets in neuroinflammation-driven chronic pain. *Nature Reviews. Drug Discovery, 13*(7), 533–548. https://doi.org/10.1038/nrd4334
11. Jäntti, H., Jonk, S., Gómez Budia, M., et al. (2024). Particulate matter from car exhaust alters function of human iPSC-derived microglia. *Particle and Fibre Toxicology, 21*(1), 6. https://doi.org/10.1186/s12989-024-00564-y
12. Jayaraj, R. L., Rodriguez, E. A., Wang, Y., & Block, M. L. (2017). Outdoor ambient air pollution and neurodegenerative diseases: The neuroinflammation hypothesis. *Current Environmental Health Reports, 4*(2), 166–179. https://doi.org/10.1007/s40572-017-0142-3
13. Levesque, S., Taetzsch, T., Lull, M. E., Johnson, J. A., McGraw, C., & Block, M. L. (2013). The role of MAC 1 in diesel exhaust particle-induced microglial activation and loss of dopaminergic neuron function. *Journal of Neurochemistry, 125*(5), 756–765. https://doi.org/10.1111/jnc.12231
14. Genc, S., Zadeoglulari, Z., Fuss, S. H., & Genc, K. (2012). The adverse effects of air pollution on the nervous system. *Journal of Toxicology, 2012*, 1–23. https://doi.org/10.1155/2012/782462
15. Yan, W., Yun, Y., Ku, T., Li, G., & Sang, N. (2016). NO2 inhalation promotes Alzheimer's disease-like progression: Cyclooxygenase-2-derived prostaglandin E2 modulation and monoacylglycerol lipase inhibition-targeted medication. *Scientific Reports, 6*(1), 22429. https://doi.org/10.1038/srep22429
16. Siracusa, R., Fusco, R., & Cuzzocrea, S. (2019). Astrocytes: Role and functions in brain pathologies. *Frontiers in Pharmacology, 10*, 1114. https://doi.org/10.3389/fphar.2019.01114
17. Gómez-Budia, M., Konttinen, H., Saveleva, L., et al. (2020). Glial smog: Interplay between air pollution and astrocyte-microglia interactions. *Neurochemistry International, 136*, 104715. https://doi.org/10.1016/j.neuint.2020.104715
18. Sies, H. (2015). Oxidative stress: A concept in redox biology and medicine. *Redox Biology, 4*, 180–183. https://doi.org/10.1016/j.redox.2015.01.002
19. Costa, L. D. O., Brandão, E., & Segundo, L. (2020). Update on epilepsy: Literature review. *Revista de Medicina* (Query date: 2023-09-22 11:16:22). https://www.revistas.usp.br/revistadc/article/download/157412/160307/403010

20. Hahad, O., Lelieveld, J., Birklein, F., Lieb, K., Daiber, A., & Münzel, T. (2020). Ambient air pollution increases the risk of cerebrovascular and neuropsychiatric disorders through induction of inflammation and oxidative stress. *International Journal of Molecular Sciences, 21*(12), 4306. https://doi.org/10.3390/ijms21124306
21. Zádori, D., Veres, G., Szalárdy, L., Klivényi, P., & Vécsei, L. (2018). Alzheimer's disease: Recent concepts on the relation of mitochondrial disturbances, excitotoxicity, neuroinflammation, and kynurenines. *Journal of Alzheimer's Disease, 62*(2), 523–547. https://doi.org/10.3233/JAD-170929
22. Breton, C. V., Song, A. Y., Xiao, J., et al. (2019). Effects of air pollution on mitochondrial function, mitochondrial DNA methylation, and mitochondrial peptide expression. *Mitochondrion, 46*, 22–29. https://doi.org/10.1016/j.mito.2019.04.001
23. Johnson, J., Mercado-Ayon, E., Mercado-Ayon, Y., et al. (2021). Mitochondrial dysfunction in the development and progression of neurodegenerative diseases. *Archives of Biochemistry and Biophysics, 702*, 108698. https://doi.org/10.1016/j.abb.2020.108698
24. Desler, C., Hansen, T. L., Frederiksen, J. B., Marcker, M. L., Singh, K. K., & Juel Rasmussen, L. (2012). Is there a link between mitochondrial reserve respiratory capacity and aging? *Journal of Aging Research, 2012*, 1–9. https://doi.org/10.1155/2012/192503
25. Ren, X., Zou, L., Zhang, X., et al. (2017). Redox signaling mediated by thioredoxin and glutathione systems in the central nervous system. *Antioxidants & Redox Signaling, 27*(13), 989–1010. https://doi.org/10.1089/ars.2016.6925
26. Chew, S., Lampinen, R., Saveleva, L., et al. (2020). Urban air particulate matter induces mitochondrial dysfunction in human olfactory mucosal cells. *Particle and Fibre Toxicology, 17*(1), 18. https://doi.org/10.1186/s12989-020-00352-4
27. Hu, R., Xie, X. Y., Xu, S. K., et al. (2017). PM2.5 exposure elicits oxidative stress responses and mitochondrial apoptosis pathway activation in HaCaT keratinocytes. *Chinese Medical Journal, 130*(18), 2205–2214. https://doi.org/10.4103/0366-6999.212942
28. Mussalo, L., Lampinen, R., Avesani, S., et al. (2024). Traffic-related ultrafine particles impair mitochondrial functions in human olfactory mucosa cells—Implications for Alzheimer's disease. *Redox Biology, 75*, 103272. https://doi.org/10.1016/j.redox.2024.103272
29. Wang, Y., Zhang, M., Li, Z., et al. (2019). Fine particulate matter induces mitochondrial dysfunction and oxidative stress in human SH-SY5Y cells. *Chemosphere, 218*, 577–588. https://doi.org/10.1016/j.chemosphere.2018.11.149
30. Murphy, M. P., Bayir, H., Belousov, V., et al. (2022). Guidelines for measuring reactive oxygen species and oxidative damage in cells and in vivo. *Nature Metabolism, 4*(6), 651–662. https://doi.org/10.1038/s42255-022-00591-z
31. Avila-Costa, M. R., Colín-Barenque, L., Fortoul, T. I., et al. (2001). Motor impairments in an oxidative stress model and its correlation with cytological changes on rat striatum and prefrontal cortex. *The International Journal of Neuroscience, 108*(3–4), 193–200. https://doi.org/10.3109/00207450108986514
32. Salvi, A., Liu, H., & Salim, S. (2020). Involvement of oxidative stress and mitochondrial mechanisms in air pollution-related neurobiological impairments. *Neurobiology of Stress, 12*, 100205. https://doi.org/10.1016/j.ynstr.2019.100205
33. Gunawan, C., Fleming, C., Irga, P. J., et al. (2024). Neurodegenerative effects of air pollutant particles: Biological mechanisms implicated for Early-Onset Alzheimer's disease. *Environment International, 185*, 108512. https://doi.org/10.1016/j.envint.2024.108512
34. Park, S. J., Lee, J., Lee, S., et al. (2020). Exposure of ultrafine particulate matter causes glutathione redox imbalance in the hippocampus: A neurometabolic susceptibility to Alzheimer's pathology. *Science of the Total Environment, 718*, 137267. https://doi.org/10.1016/j.scitotenv.2020.137267
35. Ehsanifar, M., Montazeri, Z., Taheri, M. A., Rafati, M., Behjati, M., & Karimian, M. (2021). Hippocampal inflammation and oxidative stress following exposure to diesel exhaust nanoparticles in male and female mice. *Neurochemistry International, 145*, 104989. https://doi.org/10.1016/j.neuint.2021.104989

36. Ha, S. (2021). Air pollution and neurological development in children. *Developmental Medicine and Child Neurology, 63*(4), 374–381. https://doi.org/10.1111/dmcn.14758
37. Costa, L. G., Chang, Y. C., & Cole, T. B. (2017). Developmental neurotoxicity of traffic-related air pollution: Focus on autism. *Current Environmental Health Reports, 4*(2), 156–165. https://doi.org/10.1007/s40572-017-0135-2
38. Wei, Q., Ji, Y., Gao, H., et al. (2022). Oxidative stress-mediated particulate matter affects the risk of relapse in schizophrenia patients: Air purification intervention-based panel study. *Environmental Pollution, 292*, 118348. https://doi.org/10.1016/j.envpol.2021.118348
39. Calderón-Garcidueñas, L., Solt, A. C., Henríquez-Roldán, C., et al. (2008). Long-term air pollution exposure is associated with neuroinflammation, an altered innate immune response, disruption of the blood-brain barrier, ultrafine particulate deposition, and accumulation of amyloid β-42 and α-synuclein in children and young adults. *Toxicologic Pathology, 36*(2), 289–310. https://doi.org/10.1177/0192623307313011
40. Calderón-Garcidueñas, L., Kavanaugh, M., Block, M., et al. (2012). Neuroinflammation, hyperphosphorylated tau, diffuse amyloid plaques, and down-regulation of the cellular prion protein in air pollution exposed children and young adults. *Journal of Alzheimer's Disease, 28*(1), 93–107. https://doi.org/10.3233/JAD-2011-110722
41. Kirschvink, J. L., Kobayashi-Kirschvink, A., & Woodford, B. J. (1992). Magnetite biomineralization in the human brain. *Proceedings of the National Academy of Sciences, 89*(16), 7683–7687. https://doi.org/10.1073/pnas.89.16.7683
42. Hautot, D., Pankhurst, Q. A., Khan, N., & Dobson, J. (2003). Preliminary evaluation of nanoscale biogenic magnetite in Alzheimer's disease brain tissue. *Proceedings of the Royal Society of London Series B: Biological Sciences, 270*(suppl_1). https://doi.org/10.1098/rsbl.2003.0012
43. Pankhurst, Q., Hautot, D., Khan, N., & Dobson, J. (2008). Increased levels of magnetic iron compounds in Alzheimer's disease. *Journal of Alzheimer's Disease, 13*(1), 49–52. https://doi.org/10.3233/JAD-2008-13105
44. Maher, B. A., Ahmed, I. A. M., Karloukovski, V., et al. (2016). Magnetite pollution nanoparticles in the human brain. *Proceedings of the National Academy of Sciences, 113*(39), 10797–10801. https://doi.org/10.1073/pnas.1605941113
45. Calderon-Garciduenas, L., Maronpot, R. R., Torres-Jardon, R., et al. (2003). DNA damage in nasal and brain tissues of canines exposed to air pollutants is associated with evidence of chronic brain inflammation and neurodegeneration. *Toxicologic Pathology, 31*(5), 524–538. https://doi.org/10.1080/01926230390226645
46. Ljubimova, J. Y., Braubach, O., Patil, R., et al. (2018). Coarse particulate matter (PM2.5–10) in Los Angeles Basin air induces expression of inflammation and cancer biomarkers in rat brains. *Scientific Reports, 8*(1), 5708. https://doi.org/10.1038/s41598-018-23885-3
47. Kang, Y. J., Tan, H., Lee, C. Y., & Cho, H. (2021). An air particulate pollutant induces neuroinflammation and neurodegeneration in human brain models. *Advancement of Science, 8*(21), 2101251. https://doi.org/10.1002/advs.202101251
48. Vanbrabant, K., Van Dam, D., Bongaerts, E., et al. (2024). Accumulation of ambient black carbon particles within key memory-related brain regions. *JAMA Network Open, 7*(4), e245678. https://doi.org/10.1001/jamanetworkopen.2024.5678
49. González-Maciel, A., Reynoso-Robles, R., Torres-Jardón, R., Mukherjee, P. S., & Calderón-Garcidueñas, L. (2017). Combustion-derived nanoparticles in key brain target cells and organelles in young urbanites: Culprit hidden in plain sight in Alzheimer's disease development. *Journal of Alzheimer's Disease, 59*(1), 189–208. https://doi.org/10.3233/JAD-170012
50. Nemmar, A., Holme, J. A., Rosas, I., Schwarze, P. E., & Alfaro-Moreno, E. (2013). Recent advances in particulate matter and nanoparticle toxicology: A review of the *in vivo* and *in vitro* studies. *BioMed Research International, 2013*, 1–22. https://doi.org/10.1155/2013/279371
51. Lazaridis, M. (2023). Modelling approaches to particle deposition and clearance in the human respiratory tract. *Air Quality, Atmosphere and Health, 16*(10), 1989–2002. https://doi.org/10.1007/s11869-023-01386-1

52. Wang, Y., Xiong, L., & Tang, M. (2017). Toxicity of inhaled particulate matter on the central nervous system: Neuroinflammation, neuropsychological effects and neurodegenerative disease: Inhaled PM on the CNS. *Journal of Applied Toxicology, 37*(6), 644–667. https://doi.org/10.1002/jat.3451
53. Gehr, P., & Heyder, J. (2000). Particle deposition in the respiratory tract. In *Particle-lung interactions* (pp. 252–313). CRC Press. https://doi.org/10.1201/b14423-9
54. Maynard, R., & Atkinson, R. (2024). Deposition of particles in the respiratory tract. In *Basic mathematics for students of air pollutants*. Royal Society of Chemistry. https://doi.org/10.1039/9781837672790-00085
55. Cheng, Y. S. (2003). Aerosol deposition in the extrathoracic region. *Aerosol Science and Technology, 37*(8), 659–671. https://doi.org/10.1080/02786820300906
56. Garcia, G. J. M., Schroeter, J. D., & Kimbell, J. S. (2015). Olfactory deposition of inhaled nanoparticles in humans. *Inhalation Toxicology, 27*(8), 394–403. https://doi.org/10.3109/08958378.2015.1066904
57. Schraufnagel, D. E. (2020). The health effects of ultrafine particles. *Experimental & Molecular Medicine, 52*(3), 311–317. https://doi.org/10.1038/s12276-020-0403-3
58. Jankowska-Kieltyka, M., Roman, A., & Nalepa, I. (2021). The air we breathe: Air pollution as a prevalent proinflammatory stimulus contributing to neurodegeneration. *Frontiers in Cellular Neuroscience, 15*, 647643. https://doi.org/10.3389/fncel.2021.647643
59. Leikauf, G. D., Kim, S. H., & Jang, A. S. (2020). Mechanisms of ultrafine particle-induced respiratory health effects. *Experimental & Molecular Medicine, 52*(3), 329–337. https://doi.org/10.1038/s12276-020-0394-0
60. Qi, Y., Wei, S., Xin, T., et al. (2022). Passage of exogeneous fine particles from the lung into the brain in humans and animals. *Proceedings of the National Academy of Sciences, 119*(26), e2117083119. https://doi.org/10.1073/pnas.2117083119
61. Eren, F., & Ozturk, S. (2022). Evaluation of the effect of air pollution on cognitive functions, cognitive decline, and dementia. *Annals of Indian Academy of Neurology, 25*(Suppl 1), S9–S14. https://doi.org/10.4103/aian.aian_453_22
62. Zihni, C., Mills, C., Matter, K., & Balda, M. S. (2016). Tight junctions: From simple barriers to multifunctional molecular gates. *Nature Reviews. Molecular Cell Biology, 17*(9), 564–580. https://doi.org/10.1038/nrm.2016.80
63. Belardi, B., Hamkins-Indik, T., Harris, A. R., Kim, J., Xu, K., & Fletcher, D. A. (2020). A weak link with actin organizes tight junctions to control epithelial permeability. *Developmental Cell, 54*(6), 792–804.e7. https://doi.org/10.1016/j.devcel.2020.07.022
64. Mammoto, A., Mammoto, T., Kanapathipillai, M., et al. (2013). Control of lung vascular permeability and endotoxin-induced pulmonary oedema by changes in extracellular matrix mechanics. *Nature Communications, 4*(1), 1759. https://doi.org/10.1038/ncomms2774
65. Fukuoka, A., Matsushita, K., Morikawa, T., Takano, H., & Yoshimoto, T. (2016). Diesel exhaust particles exacerbate allergic rhinitis in mice by disrupting the nasal epithelial barrier. *Clinical and Experimental Allergy, 46*(1), 142–152. https://doi.org/10.1111/cea.12597
66. Zhao, R., Guo, Z., Zhang, R., et al. (2018). Nasal epithelial barrier disruption by particulate matter ≤2.5 μm via tight junction protein degradation. *Journal of Applied Toxicology, 38*(5), 678–687. https://doi.org/10.1002/jat.3573
67. Park, S. K., Yeon, S. H., Choi, M. R., et al. (2021). Urban particulate matters may affect endoplasmic reticulum stress and tight junction disruption in nasal epithelial cells. *American Journal of Rhinology & Allergy, 35*(6), 817–829. https://doi.org/10.1177/19458924211004006
68. Lee, D. C., Choi, H., Oh, J. M., et al. (2020). Urban particulate matter regulates tight junction proteins by inducing oxidative stress via the Akt signal pathway in human nasal epithelial cells. *Toxicology Letters, 333*, 33–41. https://doi.org/10.1016/j.toxlet.2020.07.017
69. Xian, M., Ma, S., Wang, K., et al. (2020). Particulate matter 2.5 causes deficiency in barrier integrity in human nasal epithelial cells. *Allergy, Asthma & Immunology Research, 12*(1), 56. https://doi.org/10.4168/aair.2020.12.1.56

70. Paplinska-Goryca, M., Misiukiewicz-Stepien, P., Proboszcz, M., et al. (2021). Interactions of nasal epithelium with macrophages and dendritic cells variously alter urban PM-induced inflammation in healthy, asthma and COPD. *Scientific Reports, 11*(1), 13259. https://doi.org/10.1038/s41598-021-92626-w
71. Byun, J., Song, B., Lee, K., et al. (2019). Identification of urban particulate matter-induced disruption of human respiratory mucosa integrity using whole transcriptome analysis and organ-on-a chip. *Journal of Biological Engineering, 13*(1), 88. https://doi.org/10.1186/s13036-019-0219-7
72. Mussalo, L., Avesani, S., Shahbaz, M. A., et al. (2023). Emissions from modern engines induce distinct effects in human olfactory mucosa cells, depending on fuel and after-treatment. *Science of the Total Environment, 905*, 167038. https://doi.org/10.1016/j.scitotenv.2023.167038
73. Kim, S. J., Kim, N., Park, S. H., et al. (2020). Genomic approach to explore altered signaling networks of olfaction in response to diesel exhaust particles in mice. *Scientific Reports, 10*(1), 16972. https://doi.org/10.1038/s41598-020-74109-6
74. Kim, B. Y., Park, J. Y., Cho, K. J., & Bae, J. H. (2022). Effects of urban particulate matter on the olfactory system in a mouse model. *American Journal of Rhinology & Allergy, 36*(1), 81–90. https://doi.org/10.1177/19458924211026416
75. Hasegawa, Y., Namkung, H., Smith, A., et al. (2021). Causal impact of local inflammation in the nasal cavity on higher brain function and cognition. *Neuroscience Research, 172*, 110–115. https://doi.org/10.1016/j.neures.2021.04.009
76. Olesiejuk, K., & Chałubiński, M. (2023). How does particulate air pollution affect barrier functions and inflammatory activity of lung vascular endothelium? *Allergy, 78*(3), 629–638. https://doi.org/10.1111/all.15630
77. Song, C., Liu, L., Chen, J., et al. (2020). Evidence for the critical role of the PI3K signaling pathway in particulate matter-induced dysregulation of the inflammatory mediators COX-2/PGE2 and the associated epithelial barrier protein Filaggrin in the bronchial epithelium. *Cell Biology and Toxicology, 36*(4), 301–313. https://doi.org/10.1007/s10565-019-09508-1
78. Zarcone, M. C., Duistermaat, E., Van Schadewijk, A., Jedynska, A., Hiemstra, P. S., & Kooter, I. M. (2016). Cellular response of mucociliary differentiated primary bronchial epithelial cells to diesel exhaust. *American Journal of Physiology Lung Cellular and Molecular Physiology, 311*(1), L111–L123. https://doi.org/10.1152/ajplung.00064.2016
79. Park, E., Kim, B. Y., Lee, S., et al. (2024). Diesel exhaust particle exposure exacerbates ciliary and epithelial barrier dysfunction in the multiciliated bronchial epithelium models. *Ecotoxicology and Environmental Safety, 273*, 116090. https://doi.org/10.1016/j.ecoenv.2024.116090
80. Aghapour, M., Ubags, N. D., Bruder, D., et al. (2022). Role of air pollutants in airway epithelial barrier dysfunction in asthma and COPD. *European Respiratory Review, 31*(163), 210112. https://doi.org/10.1183/16000617.0112-2021
81. Sweeney, M. D., Sagare, A. P., & Zlokovic, B. V. (2018). Blood–brain barrier breakdown in Alzheimer disease and other neurodegenerative disorders. *Nature Reviews. Neurology, 14*(3), 133–150. https://doi.org/10.1038/nrneurol.2017.188
82. Kadry, H., Noorani, B., & Cucullo, L. (2020). A blood–brain barrier overview on structure, function, impairment, and biomarkers of integrity. *Fluids and Barriers of the CNS, 17*(1), 69. https://doi.org/10.1186/s12987-020-00230-3
83. Ghersi-egea, J. F., Leininger-muller, B., Cecchelli, R., & Fenstermacher, J. D. (1995). Blood-brain interfaces: Relevance to cerebral drug metabolism. *Toxicology Letters, 82-83*, 645–653. https://doi.org/10.1016/0378-4274(95)03510-9
84. Begley, D. (2004). ABC transporters and the blood-brain barrier. *Current Pharmaceutical Design, 10*(12), 1295–1312. https://doi.org/10.2174/1381612043384844
85. Geier, E. G., Chen, E. C., Webb, A., et al. (2013). Profiling solute carrier transporters in the human blood–brain barrier. *Clinical Pharmacology and Therapeutics, 94*(6), 636–639. https://doi.org/10.1038/clpt.2013.175

86. Segura-Collar, B., Mata-Martínez, P., Hernández-Laín, A., Sánchez-Gómez, P., & Gargini, R. (2022). Blood-brain barrier disruption: A common driver of central nervous system diseases. *The Neuroscientist., 28*(3), 222–237. https://doi.org/10.1177/1073858420985838
87. Hersh, A. M., Alomari, S., & Tyler, B. M. (2022). Crossing the blood-brain barrier: Advances in nanoparticle technology for drug delivery in neuro-oncology. *International Journal of Molecular Sciences, 23*(8), 4153. https://doi.org/10.3390/ijms23084153
88. Lu, D., Luo, Q., Chen, R., et al. (2020). Chemical multi-fingerprinting of exogenous ultrafine particles in human serum and pleural effusion. *Nature Communications, 11*(1), 2567. https://doi.org/10.1038/s41467-020-16427-x
89. Kreyling, W. G. (2016). Discovery of unique and ENM—Specific pathophysiologic pathways: Comparison of the translocation of inhaled iridium nanoparticles from nasal epithelium versus alveolar epithelium towards the brain of rats. *Toxicology and Applied Pharmacology, 299*, 41–46. https://doi.org/10.1016/j.taap.2016.02.004
90. Calderón-Garcidueñas, L., Vojdani, A., Blaurock-Busch, E., et al. (2014). Air pollution and children: Neural and tight junction antibodies and combustion metals, the role of barrier breakdown and brain immunity in neurodegeneration. *Journal of Alzheimer's Disease, 43*(3), 1039–1058. https://doi.org/10.3233/JAD-141365
91. Oppenheim, H. A., Lucero, J., Guyot, A. C., et al. (2013). Exposure to vehicle emissions results in altered blood brain barrier permeability and expression of matrix metalloproteinases and tight junction proteins in mice. *Particle and Fibre Toxicology, 10*(1), 62. https://doi.org/10.1186/1743-8977-10-62
92. Suwannasual, U., Lucero, J., McDonald, J. D., & Lund, A. K. (2018). Exposure to traffic-generated air pollutants mediates alterations in brain microvascular integrity in wildtype mice on a high-fat diet. *Environmental Research, 160*, 449–461. https://doi.org/10.1016/j.envres.2017.10.029
93. Suwannasual, U., Lucero, J., Davis, G., McDonald, J. D., & Lund, A. K. (2019). Mixed vehicle emissions induces angiotensin II and cerebral microvascular angiotensin receptor expression in C57Bl/6 Mice and promotes alterations in integrity in a blood-brain barrier coculture model. *Toxicological Sciences, 170*(2), 525–535. https://doi.org/10.1093/toxsci/kfz121
94. Adivi, A., Lucero, J., Simpson, N., McDonald, J. D., & Lund, A. K. (2021). Exposure to traffic-generated air pollution promotes alterations in the integrity of the brain microvasculature and inflammation in female ApoE−/− mice. *Toxicology Letters, 339*, 39–50. https://doi.org/10.1016/j.toxlet.2020.12.016
95. Woodward, N. C., Haghani, A., Johnson, R. G., et al. (2018). Prenatal and early life exposure to air pollution induced hippocampal vascular leakage and impaired neurogenesis in association with behavioral deficits. *Translational Psychiatry, 8*(1), 261. https://doi.org/10.1038/s41398-018-0317-1
96. Huuskonen, M. T., Liu, Q., Lamorie-Foote, K., et al. (2021). Air pollution particulate matter amplifies white matter vascular pathology and demyelination caused by hypoperfusion. *Frontiers in Immunology, 12*, 785519. https://doi.org/10.3389/fimmu.2021.785519
97. Boyes, W. K., & Van Thriel, C. (2020). Neurotoxicology of nanomaterials. *Chemical Research in Toxicology, 33*(5), 1121–1144. https://doi.org/10.1021/acs.chemrestox.0c00050
98. Miller, M. R., Raftis, J. B., Langrish, J. P., et al. (2017). Inhaled nanoparticles accumulate at sites of vascular disease. *ACS Nano, 11*(5), 4542–4552. https://doi.org/10.1021/acsnano.6b08551
99. Damiao Gouveia, A. C., Skovman, A., Jensen, A., et al. (2018). Telomere shortening and aortic plaque progression in Apoliprotein E knockout mice after pulmonary exposure to candle light combustion particles. *Mutagenesis, 33*(3), 253–261. https://doi.org/10.1093/mutage/gey015
100. Pele, L. C., Thoree, V., Bruggraber, S. F., et al. (2015). Pharmaceutical/food grade titanium dioxide particles are absorbed into the bloodstream of human volunteers. *Particle and Fibre Toxicology, 12*(1), 26. https://doi.org/10.1186/s12989-015-0101-9

101. Mutlu, E. A., Engen, P. A., Soberanes, S., et al. (2011). Particulate matter air pollution causes oxidant-mediated increase in gut permeability in mice. *Particle and Fibre Toxicology, 8*(1), 19. https://doi.org/10.1186/1743-8977-8-19
102. Salim, S. Y., Jovel, J., Wine, E., et al. (2014). Exposure to ingested airborne pollutant particulate matter increases mucosal exposure to bacteria and induces early onset of inflammation in neonatal IL-10–deficient mice. *Inflammatory Bowel Diseases, 20*(7), 1129–1138. https://doi.org/10.1097/MIB.0000000000000066
103. Zhou, Y., Xu, B., Wang, L., Sun, Q., Zhang, C., & Li, S. (2024). Effects of inhaled fine particulate matter on the lung injury as well as gut microbiota in broilers. *Poultry Science, 103*(4), 103426. https://doi.org/10.1016/j.psj.2024.103426
104. You, R., Ho, Y. S., & Chang, R. C. C. (2022). The pathogenic effects of particulate matter on neurodegeneration: A review. *Journal of Biomedical Science, 29*(1), 15. https://doi.org/10.1186/s12929-022-00799-x
105. Kanninen, K. M., Lampinen, R., Rantanen, L. M., et al. (2020). Olfactory cell cultures to investigate health effects of air pollution exposure: Implications for neurodegeneration. *Neurochemistry International, 136*, 104729. https://doi.org/10.1016/j.neuint.2020.104729
106. Buck, L., & Axel, R. (1991). A novel multigene family may encode odorant receptors: A molecular basis for odor recognition. *Cell, 65*(1), 175–187. https://doi.org/10.1016/0092-8674(91)90418-X
107. Genter, M. B., & Doty, R. L. (2019). Toxic exposures and the senses of taste and smell. In *Handbook of clinical neurology* (Vol. 164, pp. 389–408). Elsevier. https://doi.org/10.1016/B978-0-444-63855-7.00022-8
108. Oberdörster, G., Sharp, Z., Atudorei, V., et al. (2004). Translocation of inhaled ultrafine particles to the brain. *Inhalation Toxicology, 16*(6–7), 437–445. https://doi.org/10.1080/08958370490439597
109. Elder, A., Gelein, R., Silva, V., et al. (2006). Translocation of inhaled ultrafine manganese oxide particles to the centralnervous system. *Environmental Health Perspectives, 114*(8), 1172–1178. https://doi.org/10.1289/ehp.9030
110. Cheng, H., Saffari, A., Sioutas, C., Forman, H. J., Morgan, T. E., & Finch, C. E. (2016). Nanoscale particulate matter from urban traffic rapidly induces oxidative stress and inflammation in olfactory epithelium with concomitant effects on brain. *Environmental Health Perspectives, 124*(10), 1537–1546. https://doi.org/10.1289/EHP134
111. Hopkins, L. E., Laing, E. A., Peake, J. L., et al. (2018). Repeated iron–soot exposure and nose-to-brain transport of inhaled ultrafine particles. *Toxicologic Pathology, 46*(1), 75–84. https://doi.org/10.1177/0192623317729222
112. Liang, C., Jiang, Y., Zhang, T., et al. (2023). Atmospheric particulate matter impairs cognition by modulating synaptic function via the nose-to-brain route. *Science of the Total Environment, 857*, 159600. https://doi.org/10.1016/j.scitotenv.2022.159600
113. Calderón-Garcidueñas, L., Franco-Lira, M., Henríquez-Roldán, C., et al. (2010). Urban air pollution: Influences on olfactory function and pathology in exposed children and young adults. *Experimental and Toxicologic Pathology, 62*(1), 91–102. https://doi.org/10.1016/j.etp.2009.02.117
114. Calderón-Garcidueñas, L., González-Maciel, A., Reynoso-Robles, R., et al. (2018). Alzheimer's disease and alpha-synuclein pathology in the olfactory bulbs of infants, children, teens and adults ≤ 40 years in Metropolitan Mexico City. APOE4 carriers at higher risk of suicide accelerate their olfactory bulb pathology. *Environmental Research, 166*, 348–362. https://doi.org/10.1016/j.envres.2018.06.027
115. Lin, Y., Hu, C., Chen, A., et al. (2020). Neurotoxicity of nanoparticles entering the brain via sensory nerve-to-brain pathways: Injuries and mechanisms. *Archives of Toxicology, 94*(5), 1479–1495. https://doi.org/10.1007/s00204-020-02701-w
116. Huynh, H., Upadhyay, P., Lopez, C. H., et al. (2022). Inhalation of silver silicate nanoparticles leads to transient and differential microglial activation in the rodent olfactory bulb. *Toxicologic Pathology, 50*(6), 763–775. https://doi.org/10.1177/01926233221107607

117. Lucchini, R. G., Dorman, D. C., Elder, A., & Veronesi, B. (2012). Neurological impacts from inhalation of pollutants and the nose–brain connection. *Neurotoxicology, 33*(4), 838–841. https://doi.org/10.1016/j.neuro.2011.12.001
118. Lewis, J., Bench, G., Myers, O., et al. (2005). Trigeminal uptake and clearance of inhaled manganese chloride in rats and mice. *Neurotoxicology, 26*(1), 113–123. https://doi.org/10.1016/j.neuro.2004.06.005
119. Gianutsos, G. (1997). Accumulation of manganese in rat brain following intranasal administration. *Fundamental and Applied Toxicology, 37*(2), 102–105. https://doi.org/10.1006/faat.1997.2306
120. Carabotti, M., Scirocco, A., Maselli, M. A., & Severi, C. (2015). The gut-brain axis: Interactions between enteric microbiota, central and enteric nervous systems. *Annals of Gastroenterology, 28*(2), 203–209.
121. Morais, L. H., Schreiber, H. L., & Mazmanian, S. K. (2021). The gut microbiota–brain axis in behaviour and brain disorders. *Nature Reviews. Microbiology, 19*(4), 241–255. https://doi.org/10.1038/s41579-020-00460-0
122. Han, Y., Wang, B., Gao, H., et al. (2022). Vagus nerve and underlying impact on the gut microbiota-brain axis in behavior and neurodegenerative diseases. *Journal of Inflammation Research, 15*, 6213–6230. https://doi.org/10.2147/JIR.S384949
123. De, J. R., De-Paula, V., Forlenza, A. S., & Forlenza, O. V. (2018). Relevance of gutmicrobiota in cognition, behaviour and Alzheimer's disease. *Pharmacological Research, 136*, 29–34. https://doi.org/10.1016/j.phrs.2018.07.007
124. Alkasir, R., Li, J., Li, X., Jin, M., & Zhu, B. (2017). Human gut microbiota: The links with dementia development. *Protein & Cell, 8*(2), 90–102. https://doi.org/10.1007/s13238-016-0338-6
125. Jin, J., Xu, Z., Zhang, L., et al. (2023). Gut-derived β-amyloid: Likely a centerpiece of the gut–brain axis contributing to Alzheimer's pathogenesis. *Gut Microbes, 15*(1), 2167172. https://doi.org/10.1080/19490976.2023.2167172
126. Chen, M., & Mor, D. E. (2023). Gut-to-brain α-synuclein transmission in Parkinson's disease: Evidence for prion-like mechanisms. *International Journal of Molecular Sciences, 24*(8), 7205. https://doi.org/10.3390/ijms24087205
127. Calderón-Garcidueñas, L., González-Maciel, A., Reynoso-Robles, R., et al. (2020). Quadruple abnormal protein aggregates in brainstem pathology and exogenous metal-rich magnetic nanoparticles (and engineered Ti-rich nanorods). The substantia nigrae is a very early target in young urbanites and the gastrointestinal tract a key brainstem portal. *Environmental Research, 191*, 110139. https://doi.org/10.1016/j.envres.2020.110139
128. Yuan, J., Tan, H., Cheng, Y., et al. (2023). Air particulate pollution exposure associated with impaired cognition via microbiota gut-brain axis: An evidence from rural elderly female in northwest China. *Environmental Science and Pollution Research, 31*(4), 6398–6410. https://doi.org/10.1007/s11356-023-31504-6
129. Li, T., Fang, J., Tang, S., et al. (2022). PM2.5 exposure associated with microbiota gut-brain axis: Multi-omics mechanistic implications from the BAPE study. *The Innovation., 3*(2), 100213. https://doi.org/10.1016/j.xinn.2022.100213
130. Kish, L., Hotte, N., Kaplan, G. G., et al. (2013). Environmental particulate matter induces murine intestinal inflammatory responses and alters the gut microbiome. Bereswill S, ed. *PLoS One, 8*(4), e62220. https://doi.org/10.1371/journal.pone.0062220
131. Wang, Z., Zhang, C., Huang, F., Liu, X., Wang, Z., & Yan, B. (2021). Breakthrough of ZrO2 nanoparticles into fetal brains depends on developmental stage of maternal placental barrier and fetal blood-brain-barrier. *Journal of Hazardous Materials, 402*, 123563. https://doi.org/10.1016/j.jhazmat.2020.123563
132. Zhou, Y., Ji, J., Chen, C., & Hong, F. (2019). Retardation of axonal and dendritic outgrowth is associated with the MAPK signaling pathway in offspring mice following maternal exposure to nanosized titanium dioxide. *Journal of Agricultural and Food Chemistry, 67*(9), 2709–2715. https://doi.org/10.1021/acs.jafc.8b06992

133. Morris-Schaffer, K., Merrill, A. K., Wong, C., Jew, K., Sobolewski, M., & Cory-Slechta, D. A. (2019). Limited developmental neurotoxicity from neonatal inhalation exposure to diesel exhaust particles in C57BL/6 mice. *Particle and Fibre Toxicology, 16*(1), 1. https://doi.org/10.1186/s12989-018-0287-8
134. Hawkins, S. J., Crompton, L. A., Sood, A., et al. (2018). Nanoparticle-induced neuronal toxicity across placental barriers is mediated by autophagy and dependent on astrocytes. *Nature Nanotechnology, 13*(5), 427–433. https://doi.org/10.1038/s41565-018-0085-3
135. Onoda, A., Takeda, K., & Umezawa, M. (2017). Dose-dependent induction of astrocyte activation and reactive astrogliosis in mouse brain following maternal exposure to carbon black nanoparticle. *Particle and Fibre Toxicology, 14*(1), 4. https://doi.org/10.1186/s12989-017-0184-6
136. Shimizu, M., Tainaka, H., Oba, T., Mizuo, K., Umezawa, M., & Takeda, K. (2009). Maternal exposure to nanoparticulate titanium dioxide during the prenatal period alters gene expression related to brain development in the mouse. *Particle and Fibre Toxicology, 6*(1), 20. https://doi.org/10.1186/1743-8977-6-20
137. Bongaerts, E., Nawrot, T. S., Van Pee, T., Ameloot, M., & Bové, H. (2020). Translocation of (ultra)fine particles and nanoparticles across the placenta; A systematic review on the evidence of in vitro, ex vivo, and in vivo studies. *Particle and Fibre Toxicology, 17*(1), 56. https://doi.org/10.1186/s12989-020-00386-8
138. Bové, H., Bongaerts, E., Slenders, E., et al. (2019). Ambient black carbon particles reach the fetal side of human placenta. *Nature Communications, 10*(1), 3866. https://doi.org/10.1038/s41467-019-11654-3
139. Grafmueller, S., Manser, P., Diener, L., et al. (2015). Bidirectional transfer study of polystyrene nanoparticles across the placental barrier in an *ex vivo* human placental perfusion model. *Environmental Health Perspectives, 123*(12), 1280–1286. https://doi.org/10.1289/ehp.1409271
140. Vidmar, J., Loeschner, K., Correia, M., et al. (2018). Translocation of silver nanoparticles in the *ex vivo* human placenta perfusion model characterized by single particle ICP-MS. *Nanoscale, 10*(25), 11980–11991. https://doi.org/10.1039/C8NR02096E
141. Bongaerts, E., Aengenheister, L., Dugershaw, B. B., et al. (2021). Label-free detection of uptake, accumulation, and translocation of diesel exhaust particles in ex vivo perfused human placenta. *Journal of Nanobiotechnology, 19*(1), 144. https://doi.org/10.1186/s12951-021-00886-5
142. Seltenrich, N. (2016). $PM_{2.5}$ exposure and intrauterine inflammation: A possible mechanism for preterm and underweight birth. *Environmental Health Perspectives, 124*(10). https://doi.org/10.1289/ehp.124-A190
143. Van Den Hooven, E. H., Pierik, F. H., De Kluizenaar, Y., et al. (2012). Air pollution exposure and markers of placental growth and function: The generation R study. *Environmental Health Perspectives, 120*(12), 1753–1759. https://doi.org/10.1289/ehp.1204918
144. Zhao, Y., Wang, P., Zhou, Y., et al. (2021). Prenatal fine particulate matter exposure, placental DNA methylation changes, and fetal growth. *Environment International, 147*, 106313. https://doi.org/10.1016/j.envint.2020.106313
145. Uchida, K. (2022). Waste clearance in the brain and neuroinflammation: A novel perspective on biomarker and drug target discovery in Alzheimer's disease. *Cells, 11*(5), 919. https://doi.org/10.3390/cells11050919
146. Liu, J., Guo, Y., Zhang, C., Zeng, Y., Luo, Y., & Wang, G. (2022). Clearance systems in the brain, from structure to function. *Frontiers in Cellular Neuroscience, 15*, 729706. https://doi.org/10.3389/fncel.2021.729706
147. Mendelsohn, A. R., & Larrick, J. W. (2013). Sleep facilitates clearance of metabolites from the brain: Glymphatic function in aging and neurodegenerative diseases. *Rejuvenation Research, 16*(6), 518–523. https://doi.org/10.1089/rej.2013.1530

148. Ueno, M., Chiba, Y., Murakami, R., et al. (2019). Disturbance of intracerebral fluid clearance and blood–brain barrier in vascular cognitive impairment. *International Journal of Molecular Sciences, 20*(10), 2600. https://doi.org/10.3390/ijms20102600
149. Natale, G., Limanaqi, F., Busceti, C. L., et al. (2021). Glymphatic system as a gateway to connect neurodegeneration from periphery to CNS. *Frontiers in Neuroscience, 15*, 639140. https://doi.org/10.3389/fnins.2021.639140
150. Dudvarski Stankovic, N., Teodorczyk, M., Ploen, R., Zipp, F., & Schmidt, M. H. H. (2016). Microglia–blood vessel interactions: A double-edged sword in brain pathologies. *Acta Neuropathologica (Berl)., 131*(3), 347–363. https://doi.org/10.1007/s00401-015-1524-y
151. Martikainen, M. V., Aakko-Saksa, P., Van Den Broek, L., et al. (2021). TUBE project: Transport-derived ultrafines and the brain effects. *International Journal of Environmental Research and Public Health, 19*(1), 311. https://doi.org/10.3390/ijerph19010311
152. Makri, A., & Stilianakis, N. I. (2008). Vulnerability to air pollution health effects. *International Journal of Hygiene and Environmental Health, 211*(3–4), 326–336. https://doi.org/10.1016/j.ijheh.2007.06.005
153. Kayalar, Ö., Rajabi, H., Konyalilar, N., et al. (2024). Impact of particulate air pollution on airway injury and epithelial plasticity; underlying mechanisms. *Frontiers in Immunology, 15*, 1324552. https://doi.org/10.3389/fimmu.2024.1324552
154. Morgan, Z. E. M., Bailey, M. J., Trifonova, D. I., et al. (2023). Prenatal exposure to ambient air pollution is associated with neurodevelopmental outcomes at 2 years of age. *Environmental Health, 22*(1), 11. https://doi.org/10.1186/s12940-022-00951-y
155. Brockmeyer, S., & D'Angiulli, A. (2016). How air pollution alters brain development: The role of neuroinflammation. *Translational Neuroscience, 7*(1), 24–30. https://doi.org/10.1515/tnsci-2016-0005
156. López-Granero, C., Polyanskaya, L., Ruiz-Sobremazas, D., Barrasa, A., Aschner, M., & Alique, M. (2023). Particulate matter in human elderly: Higher susceptibility to cognitive decline and age-related diseases. *Biomolecules, 14*(1), 35. https://doi.org/10.3390/biom14010035
157. Kumar Nelson, V., Jha, N. K., Nuli, M. V., et al. (2024). Unveiling the impact of aging on BBB and Alzheimer's disease: Factors and therapeutic implications. *Ageing Research Reviews, 98*, 102224. https://doi.org/10.1016/j.arr.2024.102224
158. Mussalo, L. (2025). *Schematic representation of the pathways through which pollutant particles can enter the human body and their subsequent biological impacts*. Created in BioRender. https://BioRender.com/v12e499

Chapter 4
Toxicological Mechanisms Induced by Air Pollution

Yuanting Xie and Yanyi Xu

Keywords Air pollution · Inflammation · Oxidative stress · Gut–brain axis · Epigenetic modification

4.1 Introduction

Air pollution is a complex mixture of harmful substances in the atmosphere that pose significant risks to public health. It includes a variety of pollutants, such as particulate matter (PM) and ozone, to which humans are exposed both indoors and outdoors. The relationship between air pollution and health is well-established globally, with its impact on morbidity and mortality surpassing the combined effects of other environmental factors. In this chapter, the key biological mechanisms including: (1) triggering inflammation responses, (2) inducing oxidative stress, (3) altering the microbiome, including the airway and gut microbiota, (4) activating the autonomic nervous system, (5) causing DNA damage and impairing repair mechanisms, (6) disrupting the endocrine system *via* the HPA axis and (7) inducing epigenetic changes, by which air pollution affects the brain are introduced in detail, hoping to provide critical insight into how air pollution contributes to neurological diseases and brain health (Fig. 4.1).

Y. Xie · Y. Xu (✉)
Department of Environmental Health, School of Public Health, Fudan University, Shanghai, China
e-mail: yanyi_xu@fudan.edu.cn

K. M. Kanninen, A. R. White (eds.), *Air Pollution and the Brain*,
https://doi.org/10.1007/978-3-031-99302-2_4

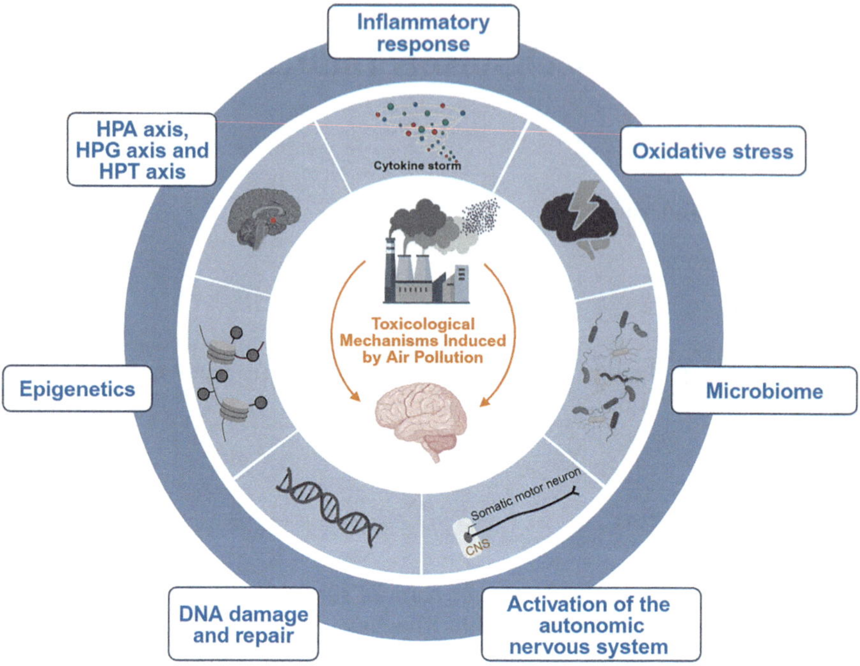

Fig. 4.1 The key biological mechanisms by which air pollution affects brain health

4.2 Inflammatory Response

4.2.1 Inflammatory Response and Air Pollution

Inflammation is a response of the immune system to harmful stimuli such as pathogens, toxic compounds, and radiation. It plays a crucial role in eliminating the adverse effects caused by these stimuli and initiating the healing process. The immune system employs a set of sensors known as damage-associated molecular patterns (DAMPs) to detect tissue injury. DAMPs are triggered in response to cell death caused by infections or toxins [1, 2]. Once activated, pattern recognition receptors (PRRs) initiate a signaling cascade that leads to the inflammatory response [3]. Air pollutant exposure has been linked to various neurological disorders, including cognitive impairment, neurodegenerative diseases such as Alzheimer's and Parkinson's disease, and stroke. A critical mechanism through which $PM_{2.5}$ contributes to these adverse health effects is the inflammatory response.

Numerous studies have demonstrated that exposure to air pollution can induce immune cell infiltration, activate inflammatory pathways, and increase expression of pro-inflammatory factors, thereby promoting adverse health outcomes [4–6]. On one hand, air pollutants may trigger a cascade of reactions through pulmonary inflammation. For example, $PM_{2.5}$ exposure is found to be capable of elevating the levels of pro-inflammatory cytokines and chemokines such as IL-6, IL-1β, and TNF-α [7]. This local inflammation in the lung then leads to systemic inflammation, in which inflammatory cells and cytokines circulate through the bloodstream and ultimately cause neuroinflammation. On the other hand, air pollutants may directly induce neuroinflammation. For instance, some studies indicated that ambient ultrafine particles may enter the brain *via* the olfactory nerve or penetrate the blood-brain barrier through the circulatory system, inducing inflammation in the central nervous system [8]. Additionally, air pollutants may also activate immune cells like microglia, polarizing them towards a pro-inflammatory state, leading to synaptic damage and cognitive impairment [9].

4.2.2 Inflammation-Related Signaling Pathways

The occurrence of inflammatory responses involves multiple signaling pathways, and a thorough investigation of the alterations in these pathways is crucial for preventing and mitigating the damage caused by air pollution through inflammation. The key signaling pathways involved in the inflammatory response include the mitogen-activated protein kinase (MAPK) signaling pathway, the NF-κB signaling pathway, the NOD-like receptor family pyrin domain containing 3 (NLRP3) signaling pathway, and the Janus kinase-signal transducer and transcription activator pathway (JAK/STAT) signaling pathway.

The MAPK cascade is a critical signaling pathway that regulates various cellular biological processes, such as proliferation, differentiation, apoptosis, and stress responses under both normal and pathological conditions. The MAPK pathway primarily includes three distinct routes: the p38 MAP kinase (p38 MAPK) pathway, the extracellular regulated protein kinase (ERK) pathway, and the c-Jun N-terminal protein kinase (JNK) pathway. Activation of these pathways requires different stimuli. Various factors, such as changes in osmotic pressure, toxins, $PM_{2.5}$, and drugs, can activate the MAPK signaling pathway. For example, experimental evidence has found that $PM_{2.5}$ can simultaneously trigger all three MAPK signaling pathways—p38 MAPK, ERK, and c-Jun pathways—thereby initiating inflammatory responses [10]. Another study demonstrated that prenatal exposure to O_3 led to cognitive deficits in male and female offsrping through the activation of MAPK and ERK signaling pathways in brain [11].

NF-κB is among the earliest transcription factors activated during the inflammatory process. In its inactive state, NF-κB is bound to IκB (inhibitor of κB) protein in the cytoplasm. When cells encounter external stimuli such as bacteria, viruses, or stress signals, the IκB kinase (IKK) complex is activated. This activation leads to the phosphorylation of IκB protein, which is subsequently ubiquitinated and degraded *via* the 26S proteasome pathway. After IκB degradation, NF-κB is released and translocated to the nucleus, where it binds to specific DNA sequences (κB sites) to activate or suppress the transcription of target genes. These genes include pro-inflammatory cytokines (such as TNF-α, IL-1β, and IL-6), immune modulators, cell adhesion molecules, and anti-apoptotic proteins, which together initiate a cascade of responses resulting in inflammatory damage [12]. Studies have shown that acute ozone exposure can activate the NF-κB signaling pathway and upregulate TLR2/4 expressions in brain [13]. $PM_{2.5}$ is also found to be capable of activating the NF-κB pathway, inducing hypothalamic inflammation and enhancing excitatory synaptic transmission in neurons [14]. In addition, NF-κB can directly or indirectly regulate oxidative stress and, in certain contexts, exert protective effects by inhibiting ROS accumulation [15]. For instance, NF-κB inhibition may lead to increased ROS production induced by TNF-α, along with heightened lipid peroxidation and protein oxidation [16].

NLRP3 inflammasome is a multiprotein complex composed of NLRP3, apoptosis-associated speck-like protein containing a CARD (ASC), and the precursor of inactive caspase-1. In its resting state, these components are present in an inactive form within the cytoplasm. The NLRP3 inflammasome can be activated by a variety of stimuli, including $PM_{2.5}$, mitochondrial damage, lysosomal instability and rupture [17, 18]. Activation of NLRP3 inflammasome is an important part of the innate immune system, and it leads to the production of active caspase-1 and interleukin-1β (IL-1β). Mature IL-1β is subsequently secreted into the extracellular space, where it participates in the inflammatory response [18, 19]. Additionally, signaling pathways like Nrf2/NLRP3 have been shown to play roles in regulating the inflammatory response caused by $PM_{2.5}$ exposure, potentially contributing to neuroinflammation and depression-like behaviors [20, 21].

The JAK/STAT pathway is a central signaling node in cellular processes, and studies have shown that it can be activated by air pollutants. After activation, the Janus kinases (JAKs), including JAK1, JAK2, JAK3, and TYK2, then phosphorylate tyrosine residues on the receptor, creating docking sites for signal transducers and activators of transcription (STATs). Once recruited, STAT proteins are phosphorylated by JAKs, causing them to dissociate from the receptor, dimerize, and translocate to the nucleus. In the nucleus, the dimerized STAT proteins bind to specific DNA sequences known as gamma-activated sites (GAS), thereby activating or repressing the transcription of target genes, including intercellular adhesion molecule 1 (ICAM-1). This JAK/STAT-mediated gene regulation impacts various cellular events, such as hematopoiesis, immune response, tissue repair, inflammation, apoptosis, and adipogenesis [22]. It has been shown that $PM_{2.5}$ promotes astrocyte activation and pro-inflammatory factor release by mediating the JAK2/STAT3 signaling pathway, which further contributes to central nervous system damage.

4.3 Oxidative Stress

4.3.1 Oxidative Stress and Air Pollution

Oxidative stress is a phenomenon caused by an imbalance between the production and accumulation of oxygen-responsive substances (ROS) in cells and tissues and the ability of biological systems to detoxify these reaction products [23]. Superoxide radicals (O_2^-), hydrogen peroxide (H_2O_2), hydroxyl radicals (^-OH), and singlet oxygen ($^1O_2$) are commonly defined as reactive oxygen species (ROS), and they are generated as metabolic by-products by biological systems [24]. Air pollution is considered as a trigger of ROS (reactive oxygen species) formation [25]. Studies show that neural and barrier damage are found in animals exposed to air pollution, with evidence in blood-brain barrier disruption, cortical neuronal degeneration, glial cell apoptosis, neurofibrillary tangles, amyloid precursor protein, and β-amyloid deposition [26].

Both *in vitro* and *in vivo* studies indicate that pollutant particles and gases can activate redox-sensitive pathways and transcription factors. Growing evidence suggests that oxidative stress in the brain plays a central role in linking air pollution to cerebrovascular and neurological diseases [27]. Studies have shown that particulate matter exposure in rats increases lipid peroxidation and oxidative stress, raises TNF-α secretion, and stimulates autophagy and intrinsic apoptosis pathways in brain tissue, ultimately contributing to cognitive impairments [28]. In addition, a synergistic mechanism has been proposed between particulate matter—which can penetrate deeply into the alveoli—and other air pollutants, including ozone, nitric oxide, soot, heavy metals, polycyclic aromatic hydrocarbons, and tobacco smoke [29]. The porous surface of airborne particles fosters an environment conducive to catalyzing reactive oxygen species (ROS) and other harmful oxidants, which may act as triggers for brain-related diseases. This synergistic enhancement of ROS formation has been experimentally confirmed [30].

4.3.2 Oxidative Stress-Related Signaling Pathways

The occurrence of oxidative stress involves multiple signaling pathways, and an in-depth study of the alteration of these signaling pathways is the key to preventing and controlling the body damage caused by air pollution. The signaling pathways involved in the development of oxidative stress include the Nrf2-antioxidant responsive element (ARE) signaling pathway, AMP-activated protein kinase (AMPK) signaling pathway, phosphoinositide-3-kinase/AKT (PI3K/AKT) signaling pathway.

The transcriptional activation of protective genes is mediated by a cis-acting element known as the antioxidant response element (ARE). The transcription factor Nrf2 binds to the ARE, stimulating the transcription of cytoprotective genes and thereby protecting cells from oxidative stress-induced cell death [31]. An increasing body of evidence indicates that the Nrf2-ARE pathway is a central intrinsic mechanism for defending against oxidative stress, playing a critical role in neutralizing ROS, free radicals, and other reactive intermediates [32]. Studies have demonstrated that the Nrf2 signaling pathway is actively involved in the oxidative stress response induced by air pollutants like $PM_{2.5}$ [33, 34]. Furthermore, research suggests that $PM_{2.5}$ exposure triggers the production of key inflammatory cytokines, such as TNF-α, IL-6, and IL-1β, which in turn stimulate Nrf2 pathway activation. $PM_{2.5}$ exposure may also upregulate inflammatory gene expression in the hypothalamus via NF-κB pathway activation, contributing to excitatory synaptic transmission in neurons.

The AMP-activated protein kinase (AMPK) complex is a heterotrimer comprising a catalytic α-subunit and regulatory β and γ subunits [35]. AMPK functions as a cellular energy sensor and plays a key role in modulating oxidative stress by regulating reactive oxygen species (ROS) release [36, 37]. Silent information regulator-1 (SIRT1), a nicotinamide adenine dinucleotide-dependent deacetylase, also contributes to oxidative stress regulation through its deacetylation activity. Together, the AMPK/SIRT1 pathway is considered an essential intracellular antioxidant mechanism [38]. Recent studies indicate that interleukin-10 (IL-10) may offer a protective effect against $PM_{2.5}$-induced damage by reducing inflammation, oxidative stress, and apoptosis through the upregulation of the AMPK/SIRT1/PGC-1α signaling pathway [39].

PI3K is an intracellular kinase that, upon activation, generates the second messenger phosphatidylinositol (3,4,5)-trisphosphate (PIP3). PIP3 binds to the pleckstrin homology (PH) domain of protein kinase B (PKB/Akt), promoting Akt phosphorylation [40]. Once activated, Akt phosphorylates mammalian target of rapamycin (mTOR) and other downstream proteins, which regulate several biological processes including oxidative stress, inflammation and autophagy [40]. Research increasingly indicates that the PI3K/Akt/mTOR signaling pathway plays a role in promoting lung cell senescence and oxidative stress [41, 42]. Moreover, studies have shown that prolonged $PM_{2.5}$ exposure in rats elevates PI3K and AKT mRNA expression in brain tissue while reducing the expression of FoxO1 mRNA. This suggests that $PM_{2.5}$-induced brain damage may be mediated by activation of the PI3K/AKT/FoxO1 signaling pathway [43].

4.4 Activation of the Autonomic Nervous System

The autonomic nervous system (ANS) consists of the sympathetic and parasympathetic branches, which innervate nearly all peripheral tissues. It receives input from the central nervous system (CNS), which integrates stimuli from the body and external environment. Specific CNS regions—including the hypothalamus, nucleus of the solitary tract, amygdala, hippocampus, and olfactory cortex—are instrumental in modulating ANS activity [44, 45]. For example, the hypothalamus plays a central role in regulating body temperature, appetite, and endocrine function, while influencing both sympathetic and parasympathetic pathways [46]. Acetylcholine and norepinephrine (or adrenaline) are the primary neurotransmitters of the ANS, acting on the parasympathetic and sympathetic systems, respectively. Over the past decade, the ANS's role in mediating health effects caused by environmental pollutants has received increased attention. Research suggests that exposure to toxins can elicit autonomic reflexes, such as sensory inputs *via* lung pathways, baroreceptor reflexes responding to blood pressure changes, or chemoreceptor reactions to variations in blood gases (pCO_2, pO_2, pH) and temperature [47]. These reflexes can result in heightened sympathetic activity or decreased parasympathetic function, ultimately leading to autonomic imbalance and adverse health outcomes [47]. Understanding the ANS's involvement in the toxicological effects of environmental pollutants is essential for elucidating the mechanisms underlying air pollution-related diseases and may provide new therapeutic targets.

Exposure to air pollution can lead to ANS dysfunction, resulting in various multisystem complications. Autonomic imbalance is thought to be a key pathway through which air pollution triggers adverse cardiovascular effects. After pollutant exposure, vagal receptors in the lungs may become stimulated, leading to autonomic dysregulation and subsequent cardiovascular impacts. This imbalance may also contribute to exacerbating brain damage. A randomized crossover study in China illustrated the cardiovascular benefits of wearing particulate-filtering respirators, noting improved heart rate variability and enhanced autonomic function [48]. Recent studies further indicate that $PM_{2.5}$ exposure alone can induce cardiac autonomic dysfunction, with ozone pre-exposure potentially intensifying this effect [49]. Alterations in cardiovascular parameters, such as blood pressure and heart rate abnormalities, can impair cerebral blood supply, affecting brain function [50, 51]. Additionally, $PM_{2.5}$ exposure may disrupt the ANS, potentially leading to arrhythmias or changes in coronary vessel tone, which could interfere with brain functions including the sleep-wake cycle and thermoregulation [52, 53].

Research into autonomic nervous control of the respiratory tract has shown that the upper respiratory tract is predominantly regulated by the trigeminal nerve, while the lower respiratory tract is controlled primarily by the vagus nerve. These nerves relay sensory information from regions such as the nasal cavity, thorax, and various visceral organs to the brain [54]. Studies indicate that ozone exposure triggers neuron activation in a time- and dose-dependent manner within the dorsolateral nucleus tractus solitarius (NTS), a region overlapping with the terminal area of vagal pulmonary afferent fibers. Sustained activation of brain areas linked to stress following ozone exposure may underlie observed neurobehavioral disturbances, including disrupted sleep, anxiety, depression, and social withdrawal. These effects likely arise from overstimulation of vagal pathways due to ozone-induced pulmonary irritation, impacting brain regions associated with stress response and emotional regulation [55, 56].

The vagus nerve, a crucial component of the ANS, plays an important role in the cholinergic anti-inflammatory pathway (CAP), which regulates the secretion of pro-inflammatory cytokines and limits the duration and intensity of inflammatory responses. Evidence suggests that exposure to $PM_{2.5}$ may disrupt ANS function by interfering with this pathway. Studies have shown that activating the CAP through vagus nerve stimulation can suppress $PM_{2.5}$-induced lung inflammation [57]. Prolonged inflammation in the lungs could damage the blood-brain barrier (BBB), potentially allowing harmful substances and pathogens to penetrate the brain, thus triggering neuroinflammation and central nervous system damage. For instance, inhaling carbon black nanoparticles has been linked to increased BBB permeability and alterations in tight junctions, leading to its disruption [58].

4.5 DNA Damage and Repair

Generally, human DNA experiences tens of thousands of damage events daily, driven by both endogenous and exogenous factors. DNA damage refers to permanent alterations in the nucleotide sequence, which can occur during replication and lead to changes in genetic traits. Damage types are typically classified as substitutions, deletions, insertions, and exon skipping. These types include point mutations, deletions (loss of nucleotides), insertions, transpositions, and double-strand breaks (DSBs). Persistent DNA damage compromises genomic stability, increasing the risk of mutations and chromosomal abnormalities. This instability can contribute to the development of various health conditions, including cancer, often manifesting years or decades after the initial exposure, imposing significant economic and social burdens [59]. Research has shown that DNA damage is a critical factor in the onset and progression of health effects caused by air pollution. For example, exposure to ultrafine particles (UFPs) has been identified as a predictor of increased purine oxidation in DNA [60]. *In vitro* studies have demonstrated that $PM_{2.5}$ exposure leads to significant DNA damage in human bronchial epithelial cells, resulting in strand breaks and altering the expression of key DNA repair genes such as XRCC1 and PARP1 [61].

DNA adducts are covalent compounds formed when electrophilic products of metabolized and activated chemical toxins bind to specific sites on DNA molecules [62]. Research has shown that various air pollutants can interact with DNA, leading to adduct formation. For example, benzo[a]pyrene (BaP), after entering the body, is metabolized by cytochrome P450 enzymes into benzo[a]pyrene diol epoxide (BPDE), an electrophilic compound considered the ultimate carcinogen of BaP in vivo. BPDE binds to DNA, forming adducts that lead to DNA damage. Moreover, exposure to other compounds, such as nitroaromatic chemicals, particulate matter, aromatic amines, and heavy metals, also induces DNA adduct formation. These adducts can cause DNA strand damage, breaks, or cross-links, disrupting normal DNA structure and function. For instance, studies have demonstrated that low concentrations of $[^{13}C_2]$-acetaldehyde can penetrate the rat brain and react with DNA to form specific adducts. This type of DNA damage may increase the risk of neurodegenerative diseases and cancer linked to air pollution. Such mechanisms highlight the potential genotoxic effects of air pollutants on the brain, where reactive compounds like acetaldehyde disrupt normal cellular processes, raising disease susceptibility [63].

DNA damage repair refers to the biological process by which cells restore damaged DNA molecules to their original structure through the activity of various enzymes. This process involves multiple pathways, including base excision repair (BER), nucleotide excision repair (NER), double-strand break repair (DSB), and mismatch repair (MMR). One key enzyme involved is oxoguanine glycosylase (OGG1), which recognizes and excises 8-oxoG (a common oxidative lesion), leaving an apurinic/apyrimidinic (AP) site that facilitates further repair. Poly ADP-ribose polymerase (PARP) and Xeroderma pigmentosum complementation group A

(XPA) are also crucial DNA repair enzymes. PARP is primarily involved in the DSB repair process, while XPA plays an essential role in the NER pathway. Studies have shown that individuals residing in heavily polluted areas, such as Mexico City, exhibit altered expression of DNA repair enzymes, particularly in the olfactory bulb and prefrontal cortex [64]. For instance, OGG1 expression decreases, while XPA expression increases in response to high pollution levels. Moreover, genetic polymorphisms in DNA repair genes may render individuals more susceptible to adverse effects from particulate matter exposure. Upon exposure to environmental toxins, cells may undergo apoptosis or activate regulatory mechanisms to arrest the cell cycle, halt DNA replication, and initiate repair pathways. This is a critical mechanism for preventing the transmission of damaged DNA to daughter cells, which could lead to mutations. Therefore, studying DNA damage repair mechanisms is fundamental for understanding the health effects of air pollution, unraveling the pathogenesis of pollution-related diseases, and developing therapeutic strategies targeting DNA repair pathways.

4.6 HPA Axis Activation

The hypothalamic-pituitary-adrenal (HPA) axis is a critical part of the neuroendocrine system, responsible for regulating the body's stress response and several physiological processes. During stress, the HPA axis is activated, starting with the paraventricular nucleus (PVN) of the hypothalamus, which releases corticotropin-releasing hormone (CRH). This signals the anterior pituitary gland to secrete adrenocorticotropic hormone (ACTH) into the bloodstream. ACTH stimulates the adrenal glands to produce cortisol, a glucocorticoid hormone that, together with epinephrine and norepinephrine from the sympathetic nervous system, prepares the body for a "fight or flight" response. This preparation involves increasing blood glucose *via* gluconeogenesis, suppressing the immune system, and enhancing fat and protein metabolism. Elevated cortisol levels are sensed by the hypothalamus and hippocampus, which regulate the stress response through a negative feedback loop [65, 66].

Air pollutants, such as $PM_{2.5}$ and ozone, can act as stressors, triggering HPA axis activation and altering stress hormone levels. While the exact mechanisms remain uncertain, previously proposed hypotheses include the activation of autonomic sensory nerves and the stimulation of stress responses in the hypothalamus *via* trigeminal and vagus nerves, which innervate the respiratory tract or through inducing the hypothalamic inflammation, as several studies have shown upregulated inflammation level in the hypothalamus by air pollutants including fine and ultrafine particulate matters. Systemic mediators like hypoxia and bioactive substances may also contribute to this process [55, 67].

Upon activation of the sympathetic nervous system, norepinephrine is released at nerve terminals via paracrine signaling, while the adrenal medulla secretes both epinephrine and norepinephrine. Concurrently, the hypothalamus responds to stress by increasing corticotropin-releasing hormone (CRH) production, prompting the pituitary gland to release adrenocorticotropic hormone (ACTH). ACTH, in turn, stimulates the adrenal cortex to synthesize and release cortisol into the bloodstream. Once released, these adrenal hormones interact with adrenergic and glucocorticoid receptors across various organs, generating systemic effects that assist the body in adapting to stressors, including inhaled pollutants [68]. Chronic activation of the HPA axis may adversely impact brain structures associated with emotion and cognition, such as the hippocampus, prefrontal cortex, and amygdala. Persistent exposure to pollutants that continually activate the HPA axis could thereby contribute to learning and memory impairments, mood disorders, and an increased risk of neurodegenerative diseases [69–71].

4.7 Microbiome

The human microbiome (HM) consists of diverse microbial communities residing in body fluids or tissues such as the skin, digestive system, biliary tract, external ear, mucous membranes, and urogenital tract [72, 73]. The HM comprises 10–100 trillion microbial cells, including bacteria, archaea, fungi, protozoa, and viruses. Gut bacteria are indispensable to host health, performing essential functions related to metabolism, shaping and supporting host immunity, and preventing pathogen invasion [74, 75]. As the "second genome" of the human body, the human microbiome plays a critical role in human health and disease development, thus it has garnered significant attentions over the past few decades [76, 77].

The gut microbiome influences multiple aspects of host physiology, and its dysbiosis can lead to a variety of diseases, ranging from localized gastrointestinal conditions to respiratory, neurological, inflammatory, liver, and cardiovascular diseases [78]. Given that inhalation is considered as the primary exposure route for atmospheric pollutants, the relationship between air pollution and gut microbiota is often overlooked. Both human and animal studies have shown that air pollutants, such as elemental carbon, black carbon, nitrates, sulfates, and toxic metals like lead and arsenic, can alter the host's gut microbiota specifically the bacterial community by influencing biofilm formation and exhibiting antimicrobial properties (Fig. 4.2) [77, 79–81]. Epidemiological research has proposed a potential correlation between air pollutants (e.g., $PM_{2.5}$, PM_{10}, nitrogen oxides) and various neurological disorders, as well as changes in brain structures. Moreover, some studies also suggest that the gut microbiome may mediate the adverse health effects such as multiple sclerosis caused by air pollution, and identified several specific gut microbial taxa of important roles in these associations [82–84].

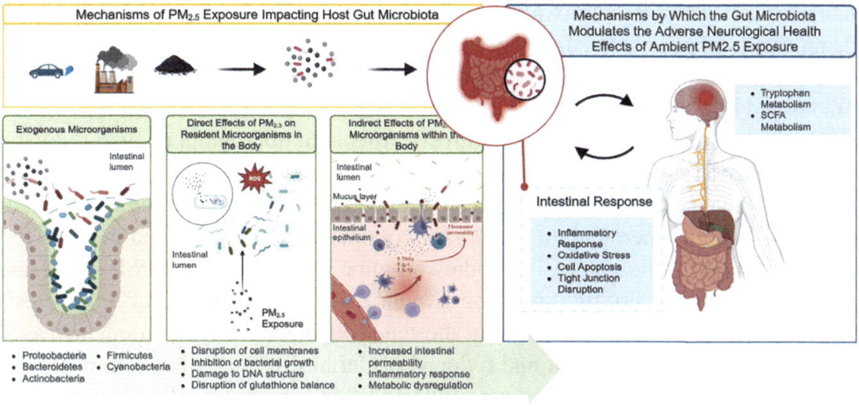

Fig. 4.2 Air pollution influences brain health *via* regulating gut microbiota

4.7.1 How Air Pollution Affects the Gut Microbiota

Air pollutants can impact gut microbiota in several ways. Take the $PM_{2.5}$ as an example. The primary components of $PM_{2.5}$ include volatile organic compounds (benzene, toluene, and polycyclic aromatic hydrocarbons), heavy metals, and soluble ions, along with microorganisms like bacteria, fungi, and viruses, which can adhere to particulate matter and enter the body. Recent studies using both culture methods [85] and non-culture techniques [86–88] have identified diverse microbial communities in atmospheric particles [89]. For example, a research based on air particulate samples from Hangzhou, China revealed that bacterial abundance in particulate matters is around 7.7×10^4 to 1.6×10^7 cells/m^3, with high abundances of phyla including *Proteobacteria*, *Cyanobacteria*, *Actinobacteria*, *Firmicutes*, and *Bacteroidetes*, along with 23 pathogenic species from 16 genera [90]. These findings suggest that $PM_{2.5}$ may influence gut microbiota by introducing exogenous microorganisms. However, current evidence on the survival and colonization of these microbes post-exposure remains limited, highlighting a need for further investigation into the composition, sources, and impacts of $PM_{2.5}$-associated microbes on the gut microbiota.

Air pollutants can also directly impact the resident gut microbiota in the body. Various body sites (such as gut, skin, mouth, and lung) harbor commensal microorganisms, with the gut microbiota dominated by phyla such as *Firmicutes*, *Bacteroidetes*, *Actinobacteria*, *Proteobacteria*, and *Fusobacteria*. The small size, large surface area, and high reactivity of ambient fine and ultrafine particles allow them to carry multiple substances, such as heavy metals and polycyclic aromatic hydrocarbons, which may interact with the gut and resident microbes upon ingestion, potentially through disrupting protein synthesis and enzymatic functions (e.g., myeloperoxidase in the gut) [91]. Additionally, heavy metals in $PM_{2.5}$, like copper, lead, and cadmium, can directly influence gut microbiota. For example, copper nanoparticles exhibit cytotoxicity, inhibiting bacterial growth and causing cell death, while copper ions can induce oxidative stress by binding to glutathione in Gram-negative bacteria, and lead and cadmium can damage bacterial cell membranes and DNA structure [92].

Furthermore, air pollution may indirectly affect gut microbiota by disrupting gut barrier function, immune response, and metabolic levels, all closely linked to gut microbial balance [93]. Air pollutants are found to be capable of reducing gut mucus thickness, increase permeability, and alter commensal microbiota and metabolites, thus impacting gut homeostasis. Studies on mice have shown that $PM_{2.5}$ exposure lowers the expression of tight junction protein ZO-1, increases colon cell apoptosis and disrupts colonic structure, causing shortened colon length. In addition, increased inflammatory cell infiltration, and reduced bacterial diversity were also observed. $PM_{2.5}$ exposure also triggers systemic inflammation, activating inflammatory pathways and elevating inflammatory markers, potentially affecting gut microbiota composition and structure [94]. In addition, $PM_{2.5}$ can alter host metabolism, affecting gut microbiota homeostasis. Metabolism is critical to physiological maintenance, and gut microbiota plays a key role in this process. For example, by using a whole-body exposure mouse model, researchers found that $PM_{2.5}$ disrupts arachidonic acid and linoleic acid metabolism [95], correlating with ileal microenvironmental changes, while levels of key metabolites involved in amino acid pathways are significantly altered in serum and bronchoalveolar lavage fluid [96].

4.7.2 *Air Pollutants Influence Brain Health Through Gut Microbiota*

By now, most studies indicate that air pollution exposure may influence brain health *by* regulating the gut microbiota-derived metabolites. Gut microbial metabolites refer to the chemical substances produced by gut bacteria during the digestion of food and metabolic processes. These metabolites include short-chain fatty acids (SCFAs), organic acids, gases, bile acids, amino acids, neurotransmitters, and vitamins, which play critical roles in maintaining gut barrier function, immune system modulation, and systemic inflammatory responses. Through the gut-brain axis, these metabolites influence mood, cognition, and behavior. Air pollutants, particularly fine particulate matter and harmful gases, may alter the composition and function of the gut microbiome, thereby affecting the production of microbial metabolites, potentially increasing the risk of neurological diseases linked to air pollution.

SCFAs, particularly, are the primary products of microbial fermentation of dietary fibers and indigestible carbohydrates in the gut [97]. Exposure to $PM_{2.5}$ has been found to increase the accumulation of SCFAs, particularly propionate, in the gut, which has various impacts on the nervous system [98, 99]. Propionate functions as a neurotransmitter, influencing neural signal transmission, and regulates the synthesis and secretion of serotonin (5-HT), a key neurotransmitter involved in mood, sleep, appetite, and cognition. Changes in serotonin levels in the gut may subsequently alter its levels in the brain. Additionally, propionate can affect dopamine levels, another critical neurotransmitter involved in motor control, reward, and pleasure perception. Alterations in the dopaminergic system due to increased propionate accumulation may be associated with neurodevelopmental disorders such as autism spectrum disorder (ASD) [100–103].

LPS, a potent immune system activator found in the outer membrane of Gram-negative bacteria, plays a key role in the mechanisms of sepsis and other inflammatory conditions. When released into circulation, LPS can cross the blood-brain barrier (BBB) and activate microglia, leading to neuroinflammation associated with various neurological disorders, including neurodegenerative diseases, depression, and anxiety. Increased levels of LPS are also linked to a higher risk of Alzheimer's disease and other forms of dementia. Thus, air pollution may negatively impact brain health by elevating intestinal levels of LPS [104, 105].

4.8 Epigenetic Modifications

Epigenetics refers to the processes by which gene function changes without altering the underlying DNA sequence. These modifications may manifest in cellular and physiological characteristics and, in addition to being part of normal development, can be triggered by environmental factors [106]. Air pollution exposure has been shown to influence epigenetic modifications that shape gene expression. Such changes can affect the development of specific diseases later in life and may be passed on to subsequent generations. The major types of epigenetic modifications include DNA methylation, histone modifications, non-coding RNA regulation, and chromatin structure remodeling [107, 108]. Emerging evidence suggests that air pollutants modulate the epigenetic landscape, from DNA methylation to miRNA expression (Fig. 4.3) [109].

DNA methylation occurs when DNA methyltransferases transfer a methyl group to the C5 position of cytosine nucleotides, forming 5-methylcytosine. Research has identified that DNA methylation, particularly in genes linked to neuroinflammation, mediates the relationship between traffic-related $PM_{2.5}$ exposure and Alzheimer's disease (AD) [110]. An animal study further shows that maternal exposure to $PM_{2.5}$ can adversely affect neurobehavior across multiple generations, leading to high methylation of the brain-derived neurotrophic factor (BDNF) promoter, which reduces BDNF expression [111]. Several mechanisms have been proposed to explain BDNF methylation changes following air pollution exposure. First, $PM_{2.5}$ exposure activates DNA methyltransferases (DNMTs) in brain regions like the hippocampus or prefrontal cortex, resulting in hypermethylation of the BDNF promoter. Second, the methyl donor S-adenosylmethionine (SAMe) can be activated by air pollutants, facilitating CpG site methylation. Third, air pollutants may suppress the expression of ten-eleven translocation (TET) family enzymes, which promote CpG demethylation [112]. Previous research has demonstrated that

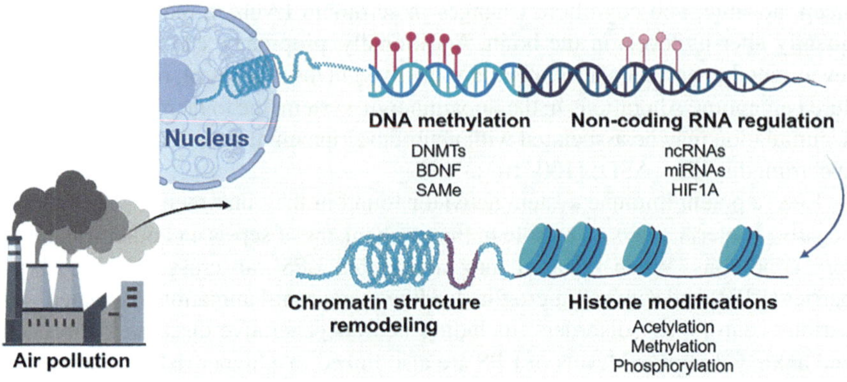

Fig. 4.3 Biological mechanisms by which air pollution affects the brain through epigenetics

hypermethylation silences the BDNF promoter, leading to decreased mRNA and protein levels of BDNF [113, 114]. BDNF is crucial for synapse growth, synaptic plasticity, and long-term potentiation in the hippocampus, which are essential for memory and cognitive function [115]. Furthermore, reduced BDNF secretion impairs neuroplasticity [116], contributing to cognitive decline and the progression of AD.

Histone modification refers to chemical modifications on histone amino acid residues, such as acetylation, methylation, and phosphorylation, which alter chromatin structure and function, thereby regulating gene expression [117–119]. In environmental contexts, air pollutants like $PM_{2.5}$ contain metals and organic compounds that can influence histone modification patterns. These changes can affect the compactness of chromatin, subsequently altering gene expression, and are linked to increased risks of neurodegenerative diseases, including AD. Specifically, air pollutants may lead to a global reduction in histone modifications, resulting in abnormal gene regulation. This process could impact brain pathologies associated with AD, such as neuroinflammation and oxidative stress. Studies suggest that pollutants, particularly iron-rich nanoparticles in $PM_{2.5}$, may reduce inhibitory histone modifications like H3K9me2/me3 in the brain, contributing to DNA damage and increasing the risk of early-onset Alzheimer's in younger individuals. These changes, observed in the brains of young people from heavily polluted cities, have been corroborated by findings in animal models, indicating that air pollution may have long-term detrimental effects on brain health [120]. Thus, alterations in histone modification may serve as a molecular mechanism linking air pollution exposure to neurodegenerative diseases.

Non-coding RNAs (ncRNAs) play a critical role in post-transcriptional regulation of gene expression through direct interactions with target genes and by coordinating responses with other epigenetic mechanisms. MicroRNAs (miRNAs), a class of small ncRNAs, can bind directly to the 3′ untranslated region (UTR), 5′ UTR, or coding sequence of mRNAs to inhibit translation or bind to promoter regions to induce transcription. miRNAs operate within tightly controlled epigenetic "circuits," influencing chromatin states to promote transcription, while also being regulated by DNA methylation and histone modifications. Early evidence shows that miRNAs respond to air pollution exposure, though the details of these underlying mechanisms remain limited [112]. For example, studies found that male C57BL/6 mice exposed to coal-burning pollutants, including sulfur dioxide, nitrogen dioxide, and $PM_{2.5}$, experienced downregulation of miR-337-5p and miR-338-5p, which may be linked to neurodegenerative changes and altered hypoxia-inducible factor 1 alpha (HIF1A) activity. HIF1A plays a crucial role in various biological processes within the nervous system. Though miRNAs have traditionally been the most studied ncRNAs in environmental health, increasing evidence indicates that other ncRNAs may also be impacted by air pollution. Recent rodent studies suggest that $PM_{2.5}$ exposure can dysregulate both mRNA and long non-coding RNA (lncRNA) expression in the hippocampus, affecting microglial function and leading to learning and memory impairments [121]. These findings underscore the urgent need for further research to explore how air pollution influences ncRNA expression, thus broadening our understanding of its effects on human health.

Chromatin remodeling refers to the dynamic structural changes within chromatin that involve compaction, decompaction, and reorganization, all of which are crucial for regulating gene expression, DNA repair, cellular differentiation, and the maintenance of cell functions [122]. Air pollutants, particularly the metals and organic compounds present in $PM_{2.5}$, can influence the activity of chromatin remodeling enzymes, leading to alterations in chromatin structure and function. These changes in chromatin remodeling may subsequently affect gene expression patterns and are linked to processes such as oxidative stress, inflammatory responses, and increased risk of neurodegenerative diseases like AD [123]. Therefore, alterations in chromatin remodeling could be a molecular mechanism through which air pollution negatively impacts neurological health.

4.9 Conclusions

Air pollution is a complex issue that significantly impacts human health, the environment, and biodiversity [124]. Research indicates that CNS is a primary target of air pollution, second only to cardiopulmonary system. Numerous epidemiological and toxicological studies have linked air pollution to CNS-related diseases, demonstrating its substantial effects on brain health through various biological mechanisms, such as inflammation, oxidative stress, microbiome dysbiosis, HPA axis activation, epigenetic modifications, DNA damage, and impaired repair. These mechanisms not only highlight the complexity of air pollution's impact on brain health but also underscore the urgency of addressing this global health challenge.

Despite significant advances in understanding the mechanisms underlying air pollution's effects on the brain, substantial knowledge gaps remain. These include the time effect of air pollution on the brain (acute *vs.* chronic), the specific role of individual air pollutant and their interactions with each other (even with other meteorological factors such as temperature and humidity) on the brain, and the discovery of key effective target molecules and development of protective strategies. Additionally, there is a need for further exploration of biomarkers to investigate early interventions to mitigate the effects of air pollution on brain health, particularly in vulnerable populations such as children and the elderly [125]. Importantly, the evidence underscores the urgent need for policies to reduce air pollution exposure and enhance public health, highlighting the critical role of interdisciplinary collaboration between environmental science, neuroscience, and policy management.

References

1. Rock, K. L., & Kono, H. (2008). The inflammatory response to cell death. *Annual Review of Pathology, 3*, 99–126. https://doi.org/10.1146/annurev.pathmechdis.3.121806.151456
2. Roh, J. S., & Sohn, D. H. (2018). Damage-associated molecular patterns in inflammatory diseases. *Immune Network, 18*(4), e27. https://doi.org/10.4110/in.2018.18.e27
3. Tang, D., Kang, R., Coyne, C. B., Zeh, H. J., & Lotze, M. T. (2012). PAMPs and DAMPs: Signal 0s that spur autophagy and immunity. *Immunological Reviews, 249*(1), 158–175. https://doi.org/10.1111/j.1600-065X.2012.01146.x
4. Liu, C., Yang, J., Guan, L., Zhu, Y., & Geng, X. (2020). Filtered air intervention reduces inflammation and hypothalamus-pituitary-adrenal axis activation in adult male and female rats after PM 2.5 exposure. *Environmental Science and Pollution Research International, 27*(28), 35341–35348. https://doi.org/10.1007/s11356-020-09564-9
5. Salim, S. Y., Jovel, J., Wine, E., Kaplan, G. G., Vincent, R., Thiesen, A., et al. (2014). Exposure to ingested airborne pollutant particulate matter increases mucosal exposure to bacteria and induces early onset of inflammation in neonatal IL-10-deficient mice. *Inflammatory Bowel Diseases, 20*(7), 1129–1138. https://doi.org/10.1097/MIB.0000000000000066
6. Pasalic, E., Hayat, M. J., & Greenwald, R. (2016). Air pollution, physical activity, and markers of acute airway oxidative stress and inflammation in adolescents. *The Journal of the Georgia Public Health Association, 6*(2 Suppl), 314–330. https://doi.org/10.21633/jgpha.6.2s19
7. Xie, Y., Tao, S., Pan, B., Yang, W., Shao, W., Fang, X. et al. (2023) Cholinergic anti-inflammatory pathway mediates diesel exhaust PM(2.5)-induced pulmonary and systemic inflammation. *Journal of Hazardous Materials, 48*, 131951. https://doi.org/10.1016/j.jhazmat.2023.131951
8. Kang, Y. J., Tan, H. Y., Lee, C. Y., & Cho, H. (2021). An air particulate pollutant induces neuroinflammation and neurodegeneration in human brain models. *Advanced Science (Weinh), 8*(21), e2101251. https://doi.org/10.1002/advs.202101251
9. Kim, R. E., Shin, C. Y., Han, S. H., & Kwon, K. J. (2020). Astaxanthin suppresses PM2.5-induced neuroinflammation by regulating akt phosphorylation in BV-2 microglial cells. *International Journal of Molecular Sciences, 21*(19), 7227. https://doi.org/10.3390/ijms21197227
10. Zhen, A. X., Piao, M. J., Hyun, Y. J., Kang, K. A., Madushan Fernando, P. D. S., Cho, S. J., et al. (2019). Diphlorethohydroxycarmalol attenuates fine particulate matter-induced subcellular skin dysfunction. *Marine Drugs, 17*(2), 95. https://doi.org/10.3390/md17020095
11. Custodio, V., Rubio, C., & Paz, C. (2019). Prenatal ozone exposure induces memory deficiencies in newborns rats. *Frontiers in Molecular Neuroscience, 12*, 244. https://doi.org/10.3389/fnmol.2019.00244
12. Lawrence, T. (2009). The nuclear factor NF-kappaB pathway in inflammation. *Cold Spring Harbor Perspectives in Biology, 1*(6), a001651. https://doi.org/10.1101/cshperspect.a001651

13. Zhu, L., Ding, S., Xu, L., & Wu, Z. (2022). Ozone treatment alleviates brain injury in cerebral ischemic rats by inhibiting the NF-kappaB signaling pathway and autophagy. *Cell Cycle, 21*(4), 406–415. https://doi.org/10.1080/15384101.2021.2020961
14. Ku, T., Li, B., Gao, R., Zhang, Y., Yan, W., Ji, X., et al. (2017). NF-kappaB-regulated microRNA-574-5p underlies synaptic and cognitive impairment in response to atmospheric PM(2.5) aspiration. *Particle and Fibre Toxicology, 14*(1), 34. https://doi.org/10.1186/s12989-017-0215-3
15. Priya Dharshini, L. C., Vishnupriya, S., Sakthivel, K. M., & Rasmi, R. R. (2020). Oxidative stress responsive transcription factors in cellular signalling transduction mechanisms. *Cellular Signalling, 72*, 109670. https://doi.org/10.1016/j.cellsig.2020.109670
16. Djavaheri-Mergny, M., Javelaud, D., Wietzerbin, J., & Besancon, F. (2004). NF-kappaB activation prevents apoptotic oxidative stress via an increase of both thioredoxin and MnSOD levels in TNFalpha-treated Ewing sarcoma cells. *FEBS Letters, 578*(1–2), 111–115. https://doi.org/10.1016/j.febslet.2004.10.082
17. Pandey, A., Shen, C., Feng, S., & Man, S. M. (2021). Cell biology of inflammasome activation. *Trends in Cell Biology, 31*(11), 924–939. https://doi.org/10.1016/j.tcb.2021.06.010
18. Zeng, X., Liu, D., Wu, W., & Huo, X. (2022). PM(2.5) exposure inducing ATP alteration links with NLRP3 inflammasome activation. *Environmental Science and Pollution Research International, 29*(17), 24445–24456. https://doi.org/10.1007/s11356-021-16405-w
19. Zheng, D., Liwinski, T., & Elinav, E. (2020). Inflammasome activation and regulation: Toward a better understanding of complex mechanisms. *Cell Discovery, 6*, 36. https://doi.org/10.1038/s41421-020-0167-x
20. Chu, C., Zhang, H., Cui, S., Han, B., Zhou, L., Zhang, N., et al. (2019). Ambient PM2.5 caused depressive-like responses through Nrf2/NLRP3 signaling pathway modulating inflammation. *Journal of Hazardous Materials, 369*, 180–190. https://doi.org/10.1016/j.jhazmat.2019.02.026
21. Shi, J. Q., Wang, B. R., Jiang, T., Gao, L., Zhang, Y. D., & Xu, J. (2020). NLRP3 inflammasome: A potential therapeutic target in fine particulate matter-induced neuroinflammation in Alzheimer's disease. *Journal of Alzheimer's Disease, 77*(3), 923–934. https://doi.org/10.3233/JAD-200359
22. Sarapultsev, A., Gusev, E., Komelkova, M., Utepova, I., Luo, S., & Hu, D. (2023). JAK-STAT signaling in inflammation and stress-related diseases: Implications for therapeutic interventions. *Molecular Biomedicine, 4*(1), 40. https://doi.org/10.1186/s43556-023-00151-1
23. Pizzino, G., Irrera, N., Cucinotta, M., Pallio, G., Mannino, F., Arcoraci, V., et al. (2017). Oxidative stress: Harms and benefits for human health. *Oxidative Medicine and Cellular Longevity, 2017*, 8416763. https://doi.org/10.1155/2017/8416763
24. Sato, H., Shibata, M., Shimizu, T., Shibata, S., Toriumi, H., Ebine, T., et al. (2013). Differential cellular localization of antioxidant enzymes in the trigeminal ganglion. *Neuroscience, 248*, 345–358. https://doi.org/10.1016/j.neuroscience.2013.06.010
25. Block, M. L., & Calderon-Garciduenas, L. (2009). Air pollution: Mechanisms of neuroinflammation and CNS disease. *Trends in Neurosciences, 32*(9), 506–516. https://doi.org/10.1016/j.tins.2009.05.009
26. Calderon-Garciduenas, L., Azzarelli, B., Acuna, H., Garcia, R., Gambling, T. M., Osnaya, N., et al. (2002). Air pollution and brain damage. *Toxicologic Pathology, 30*(3), 373–389. https://doi.org/10.1080/01926230252929954
27. Wang, Y., Xiong, L., & Tang, M. (2017). Toxicity of inhaled particulate matter on the central nervous system: Neuroinflammation, neuropsychological effects and neurodegenerative disease. *Journal of Applied Toxicology, 37*(6), 644–667. https://doi.org/10.1002/jat.3451
28. Hajipour, S., Farbood, Y., Gharib-Naseri, M. K., Goudarzi, G., Rashno, M., Maleki, H., et al. (2020). Exposure to ambient dusty particulate matter impairs spatial memory and hippocampal LTP by increasing brain inflammation and oxidative stress in rats. *Life Sciences, 242*, 117210. https://doi.org/10.1016/j.lfs.2019.117210

29. Park, J. H., Troxel, A. B., Harvey, R. G., & Penning, T. M. (2006). Polycyclic aromatic hydrocarbon (PAH) o-quinones produced by the aldo-keto-reductases (AKRs) generate abasic sites, oxidized pyrimidines, and 8-oxo-dGuo via reactive oxygen species. *Chemical Research in Toxicology, 19*(5), 719–728. https://doi.org/10.1021/tx0600245
30. Valavanidis, A., Vlachogianni, T., Fiotakis, K., & Loridas, S. (2013). Pulmonary oxidative stress, inflammation and cancer: Respirable particulate matter, fibrous dusts and ozone as major causes of lung carcinogenesis through reactive oxygen species mechanisms. *International Journal of Environmental Research and Public Health, 10*(9), 3886–3907. https://doi.org/10.3390/ijerph10093886
31. Johnson, J. A., Johnson, D. A., Kraft, A. D., Calkins, M. J., Jakel, R. J., Vargas, M. R., & Chen, P. C. (2008). The Nrf2-ARE pathway: An indicator and modulator of oxidative stress in neurodegeneration. *Annals of the New York Academy of Sciences, 1147*, 61–69. https://doi.org/10.1196/annals.1427.036
32. Shang, Y., Zhou, Q., Wang, T., Jiang, Y., Zhong, Y., Qian, G., et al. (2017). Airborne nitro-PAHs induce Nrf2/ARE defense system against oxidative stress and promote inflammatory process by activating PI3K/Akt pathway in A549 cells. *Toxicology In Vitro, 44*, 66–73. https://doi.org/10.1016/j.tiv.2017.06.017
33. Zhao, C., Pu, W., Wazir, J., Jin, X., Wei, L., Song, S., et al. (2022). Long-term exposure to PM2.5 aggravates pulmonary fibrosis and acute lung injury by disrupting Nrf2-mediated antioxidant function. *Environmental Pollution, 313*, 120017. https://doi.org/10.1016/j.envpol.2022.120017
34. Saveleva, L., Cervena, T., Mengoni, C., Sima, M., Krejcik, Z., Vrbova, K., et al. (2024). Transcriptomic and epigenomic profiling reveals altered responses to diesel emissions in Alzheimer's disease both in vitro and in population-based data. *Alzheimers Dement, 20*(12), 8825–8843. https://doi.org/10.1002/alz.14347
35. Herzig, S., & Shaw, R. J. (2018). AMPK: Guardian of metabolism and mitochondrial homeostasis. *Nature Reviews. Molecular Cell Biology, 19*(2), 121–135. https://doi.org/10.1038/nrm.2017.95
36. Auciello, F. R., Ross, F. A., Ikematsu, N., & Hardie, D. G. (2014). Oxidative stress activates AMPK in cultured cells primarily by increasing cellular AMP and/or ADP. *FEBS Letters, 588*(18), 3361–3366. https://doi.org/10.1016/j.febslet.2014.07.025
37. Han, X., Tai, H., Wang, X., Wang, Z., Zhou, J., Wei, X., et al. (2016). AMPK activation protects cells from oxidative stress-induced senescence via autophagic flux restoration and intracellular NAD(+) elevation. *Aging Cell, 15*(3), 416–427. https://doi.org/10.1111/acel.12446
38. Feng, K., Chen, Z., Pengcheng, L., Zhang, S., & Wang, X. (2019). Quercetin attenuates oxidative stress-induced apoptosis via SIRT1/AMPK-mediated inhibition of ER stress in rat chondrocytes and prevents the progression of osteoarthritis in a rat model. *Journal of Cellular Physiology, 234*(10), 18192–18205. https://doi.org/10.1002/jcp.28452
39. Zhang, N., Li, P., Lin, H., Shuo, T., Ping, F., Su, L., & Chen, G. (2021). IL-10 ameliorates PM2.5-induced lung injury by activating the AMPK/SIRT1/PGC-1alpha pathway. *Environmental Toxicology and Pharmacology, 86*, 103659. https://doi.org/10.1016/j.etap.2021.103659
40. Shiau, J. P., Chuang, Y. T., Cheng, Y. B., Tang, J. Y., Hou, M. F., Yen, C. Y., & Chang, H. W. (2022). Impacts of oxidative stress and PI3K/AKT/mTOR on metabolism and the future direction of investigating fucoidan-modulated metabolism. *Antioxidants (Basel), 11*(5), 911. https://doi.org/10.3390/antiox11050911
41. Liu, Y., Kong, H., Cai, H., Chen, G., Chen, H., & Ruan, W. (2023). Progression of the PI3K/Akt signaling pathway in chronic obstructive pulmonary disease. *Frontiers in Pharmacology, 14*, 1238782. https://doi.org/10.3389/fphar.2023.1238782
42. Xiaofei, Y., Tingting, L., Xuan, W., & Zhiyi, H. (2022). Erythromycin attenuates oxidative stress-induced cellular senescence via the PI3K-mTOR signaling pathway in chronic obstructive pulmonary disease. *Frontiers in Pharmacology, 13*, 1043474. https://doi.org/10.3389/fphar.2022.1043474

43. Song, L., Pan, K., Du, X., Jiang, S., Zeng, X., Zhang, J., et al. (2021). Ambient PM(2.5)-induced brain injury is associated with the activation of PI3K/AKT/FoxO1 pathway. *Environmental Science and Pollution Research International, 28*(48), 68276–68287. https://doi.org/10.1007/s11356-021-15405-0
44. Williams, L. M., Barton, M. J., Kemp, A. H., Liddell, B. J., Peduto, A., Gordon, E., & Bryant, R. A. (2005). Distinct amygdala-autonomic arousal profiles in response to fear signals in healthy males and females. *NeuroImage, 28*(3), 618–626. https://doi.org/10.1016/j.neuroimage.2005.06.035
45. Sun, Q., Li, X., Ren, M., Zhao, M., Zhong, Q., Ren, Y., et al. (2019). A whole-brain map of long-range inputs to GABAergic interneurons in the mouse medial prefrontal cortex. *Nature Neuroscience, 22*(8), 1357–1370. https://doi.org/10.1038/s41593-019-0429-9
46. Biddle, C. (2006). The neurobiology of the human febrile response. *AANA Journal, 74*(2), 145–150.
47. Perez, C. M., Hazari, M. S., & Farraj, A. K. (2015). Role of autonomic reflex arcs in cardiovascular responses to air pollution exposure. *Cardiovascular Toxicology, 15*(1), 69–78. https://doi.org/10.1007/s12012-014-9272-0
48. Shi, J., Lin, Z., Chen, R., Wang, C., Yang, C., Cai, J., et al. (2017). Cardiovascular benefits of wearing particulate-filtering respirators: A randomized crossover trial. *Environmental Health Perspectives, 125*(2), 175–180. https://doi.org/10.1289/EHP73
49. Wang, G., Zhen, L., Lu, P., Jiang, R., & Song, W. (2013). Effects of ozone and fine particulate matter (PM2.5) on rat cardiac autonomic nervous system and systemic inflammation. *Wei Sheng Yan Jiu, 42*(4), 554–560.
50. Drummond, J. C. (2019). Blood pressure and the brain: How low can you go? *Anesthesia and Analgesia, 128*(4), 759–771. https://doi.org/10.1213/ANE.0000000000004034
51. Prabhakaran, S. (2019). Blood pressure, brain volume and white matter hyperintensities, and dementia risk. *JAMA, 322*(6), 512–513. https://doi.org/10.1001/jama.2019.10849
52. Chadda, K. R., Ajijola, O. A., Vaseghi, M., Shivkumar, K., Huang, C. L., & Jeevaratnam, K. (2018). Ageing, the autonomic nervous system and arrhythmia: From brain to heart. *Ageing Research Reviews, 48*, 40–50. https://doi.org/10.1016/j.arr.2018.09.005
53. Taggart, P., Critchley, H., & Lambiase, P. D. (2011). Heart-brain interactions in cardiac arrhythmia. *Heart, 97*(9), 698–708. https://doi.org/10.1136/hrt.2010.209304
54. Mazzone, S. B., & Undem, B. J. (2016). Vagal afferent innervation of the airways in health and disease. *Physiological Reviews, 96*(3), 975–1024. https://doi.org/10.1152/physrev.00039.2015
55. Kodavanti, U. P., Jackson, T. W., Henriquez, A. R., Snow, S. J., Alewel, D. I., & Costa, D. L. (2023). Air pollutant impacts on the brain and neuroendocrine system with implications for peripheral organs: A perspective. *Inhalation Toxicology, 35*(3–4), 109–126. https://doi.org/10.1080/08958378.2023.2172486
56. Gackiere, F., Saliba, L., Baude, A., Bosler, O., & Strube, C. (2011). Ozone inhalation activates stress-responsive regions of the CNS. *Journal of Neurochemistry, 117*(6), 961–972. https://doi.org/10.1111/j.1471-4159.2011.07267.x
57. Xie, Y., Tao, S., Pan, B., Yang, W., Shao, W., Fang, X., et al. (2023). Cholinergic anti-inflammatory pathway mediates diesel exhaust PM(2.5)-induced pulmonary and systemic inflammation. *Journal of Hazardous Materials, 458*, 131951. https://doi.org/10.1016/j.jhazmat.2023.131951
58. Cheng, W. T., Zhang, W. J., Xia, X. W., Zhang, J. Z., Wang, M. Y., Li, Y. T., et al. (2023). The domino effect in inhaled carbon black nanoparticles triggers blood-brain barrier disruption via altering circulatory inflammation. *Nano Today, 48*, 101721. https://doi.org/10.1016/j.nantod.2022.101721
59. Kazensky, L., Matkovic, K., Geric, M., Zegura, B., Pehnec, G., & Gajski, G. (2024). Impact of indoor air pollution on DNA damage and chromosome stability: A systematic review. *Archives of Toxicology, 98*(9), 2817–2841. https://doi.org/10.1007/s00204-024-03785-4

60. Vinzents, P. S., Moller, P., Sorensen, M., Knudsen, L. E., Hertel, O., Jensen, F. P., et al. (2005). Personal exposure to ultrafine particles and oxidative DNA damage. *Environmental Health Perspectives, 113*(11), 1485–1490. https://doi.org/10.1289/ehp.7562
61. Jasso-Pineda, Y., Diaz-Barriga, F., Yanez-Estrada, L., Perez-Vazquez, F. J., & Perez-Maldonado, I. N. (2015). DNA damage in Mexican children living in high-risk contaminated scenarios. *Science of the Total Environment, 518-519*, 38–48. https://doi.org/10.1016/j.scitotenv.2015.02.073
62. Quezada-Maldonado, E. M., Sanchez-Perez, Y., Chirino, Y. I., & Garcia-Cuellar, C. M. (2021). Airborne particulate matter induces oxidative damage, DNA adduct formation and alterations in DNA repair pathways. *Environmental Pollution, 287*, 117313. https://doi.org/10.1016/j.envpol.2021.117313
63. Sanchez, A. B., Garcia, C. C. M., Freitas, F. P., Batista, G. L., Lopes, F. S., Carvalho, V. H., et al. (2018). DNA adduct formation in the lungs and brain of rats exposed to low concentrations of [(13)C(2)]-acetaldehyde. *Chemical Research in Toxicology, 31*(5), 332–339. https://doi.org/10.1021/acs.chemrestox.8b00016
64. Calderon-Garciduenas, L., Serrano-Sierra, A., Torres-Jardon, R., Zhu, H., Yuan, Y., Smith, D., et al. (2013). The impact of environmental metals in young urbanites' brains. *Experimental and Toxicologic Pathology, 65*(5), 503–511. https://doi.org/10.1016/j.etp.2012.02.006
65. de Weerth, C. (2017). Do bacteria shape our development? Crosstalk between intestinal microbiota and HPA axis. *Neuroscience and Biobehavioral Reviews, 83*, 458–471. https://doi.org/10.1016/j.neubiorev.2017.09.016
66. Kinlein, S. A., Phillips, D. J., Keller, C. R., & Karatsoreos, I. N. (2019). Role of corticosterone in altered neurobehavioral responses to acute stress in a model of compromised hypothalamic-pituitary-adrenal axis function. *Psychoneuroendocrinology, 102*, 248–255. https://doi.org/10.1016/j.psyneuen.2018.12.010
67. Verheyen, V. J., Remy, S., Lambrechts, N., Govarts, E., Colles, A., Poelmans, L., et al. (2021). Residential exposure to air pollution and access to neighborhood greenspace in relation to hair cortisol concentrations during the second and third trimester of pregnancy. *Environmental Health, 20*(1), 11. https://doi.org/10.1186/s12940-021-00697-z
68. Herman, J. P., Nawreen, N., Smail, M. A., & Cotella, E. M. (2020). Brain mechanisms of HPA axis regulation: Neurocircuitry and feedback in context Richard Kvetnansky lecture. *Stress, 23*(6), 617–632. https://doi.org/10.1080/10253890.2020.1859475
69. Du, X., & Pang, T. Y. (2015). Is dysregulation of the HPA-axis a core pathophysiology mediating Co-morbid depression in neurodegenerative diseases? *Frontiers in Psychiatry, 6*, 32. https://doi.org/10.3389/fpsyt.2015.00032
70. Frankiensztajn, L. M., Elliott, E., & Koren, O. (2020). The microbiota and the hypothalamus-pituitary-adrenocortical (HPA) axis, implications for anxiety and stress disorders. *Current Opinion in Neurobiology, 62*, 76–82. https://doi.org/10.1016/j.conb.2019.12.003
71. Zaletel, I., Filipovic, D., & Puskas, N. (2016). Chronic stress, hippocampus and parvalbumin-positive interneurons: What do we know so far? *Reviews in the Neurosciences, 27*(4), 397–409. https://doi.org/10.1515/revneuro-2015-0042
72. Ursell, L. K., Metcalf, J. L., Parfrey, L. W., & Knight, R. (2012). Defining the human microbiome. *Nutrition Reviews, 70*(Suppl 1), S38–S44. https://doi.org/10.1111/j.1753-4887.2012.00493.x
73. Mousavi, S. E., Delgado-Saborit, J. M., Adivi, A., Pauwels, S., & Godderis, L. (2022). Air pollution and endocrine disruptors induce human microbiome imbalances: A systematic review of recent evidence and possible biological mechanisms. *Science of the Total Environment, 816*, 151654. https://doi.org/10.1016/j.scitotenv.2021.151654
74. Vernocchi, P., Del Chierico, F., & Putignani, L. (2020). Gut microbiota metabolism and interaction with food components. *International Journal of Molecular Sciences, 21*(10), 3688. https://doi.org/10.3390/ijms21103688

75. Ventura, M., O'Toole, P. W., de Vos, W. M., & van Sinderen, D. (2018). Selected aspects of the human gut microbiota. *Cellular and Molecular Life Sciences, 75*(1), 81–82. https://doi.org/10.1007/s00018-017-2669-8
76. Cho, I., & Blaser, M. J. (2012). The human microbiome: At the interface of health and disease. *Nature Reviews. Genetics, 13*(4), 260–270. https://doi.org/10.1038/nrg3182
77. Van Pee, T., Nawrot, T. S., van Leeuwen, R., & Hogervorst, J. (2023). Ambient particulate air pollution and the intestinal microbiome; A systematic review of epidemiological, in vivo and, in vitro studies. *Science of the Total Environment, 878*, 162769. https://doi.org/10.1016/j.scitotenv.2023.162769
78. Durack, J., & Lynch, S. V. (2019). The gut microbiome: Relationships with disease and opportunities for therapy. *The Journal of Experimental Medicine, 216*(1), 20–40. https://doi.org/10.1084/jem.20180448
79. Adams, K., Greenbaum, D. S., Shaikh, R., van Erp, A. M., & Russell, A. G. (2015). Particulate matter components, sources, and health: Systematic approaches to testing effects. *Journal of the Air & Waste Management Association (1995), 65*(5), 544–558. https://doi.org/10.1080/10962247.2014.1001884
80. Li, X., Brejnrod, A. D., Ernst, M., Rykaer, M., Herschend, J., Olsen, N. M. C., et al. (2019). Heavy metal exposure causes changes in the metabolic health-associated gut microbiome and metabolites. *Environment International, 126*, 454–467. https://doi.org/10.1016/j.envint.2019.02.048
81. Roslund, M. I., Rantala, S., Oikarinen, S., Puhakka, R., Hui, N., Parajuli, A., et al. (2019). Endocrine disruption and commensal bacteria alteration associated with gaseous and soil PAH contamination among daycare children. *Environment International, 130*, 104894. https://doi.org/10.1016/j.envint.2019.06.004
82. Zhou, X., Guo, Z., Ling, Y., Teng, W., Cui, J., Yan, Z., et al. (2024). Causal effect of air pollution on the risk of brain health and potential mediation by gut microbiota. *Ecotoxicology and Environmental Safety, 285*, 117080. https://doi.org/10.1016/j.ecoenv.2024.117080
83. Yuan, J., Tan, H., Cheng, Y., Ma, X., Jiang, S., Hou, X., et al. (2024). Air particulate pollution exposure associated with impaired cognition via microbiota gut-brain axis: An evidence from rural elderly female in northwest China. *Environmental Science and Pollution Research International, 31*(4), 6398–6410. https://doi.org/10.1007/s11356-023-31504-6
84. Cabrera, C., Vicens, P., & Torrente, M. (2021). Modifiable risk factors for dementia: The role of gut microbiota. *Current Alzheimer Research, 18*(13), 993–1009. https://doi.org/10.2174/1567205018666211215152411
85. Alghamdi, M. A., Shamy, M., Redal, M. A., Khoder, M., Awad, A. H., & Elserougy, S. (2014). Microorganisms associated particulate matter: A preliminary study. *Science of the Total Environment, 479–480*, 109–116. https://doi.org/10.1016/j.scitotenv.2014.02.006
86. Yan, D., Zhang, T., Su, J., Zhao, L. L., Wang, H., Fang, X. M., et al. (2016). Diversity and composition of airborne fungal community associated with particulate matters in Beijing during haze and non-haze days. *Frontiers in Microbiology, 7*, 487. https://doi.org/10.3389/fmicb.2016.00487
87. Zhong, S., Zhang, L., Jiang, X., & Gao, P. (2019). Comparison of chemical composition and airborne bacterial community structure in PM2.5 during haze and non-haze days in the winter in Guilin, China. *Science of the Total Environment, 655*, 202–210. https://doi.org/10.1016/j.scitotenv.2018.11.268
88. Romano, S., Becagli, S., Lucarelli, F., Rispoli, G., & Perrone, M. R. (2020). Airborne bacteria structure and chemical composition relationships in winter and spring PM10 samples over southeastern Italy. *Science of the Total Environment, 730*, 138899. https://doi.org/10.1016/j.scitotenv.2020.138899
89. Yan, D., Zhang, T., Bai, J. L., Su, J., Zhao, L. L., Wang, H., et al. (2021). Isolation, characterization, and antimicrobial activity of bacterial and fungal representatives associated with particulate matter during haze and non-haze days. *Frontiers in Microbiology, 12*, 793037. https://doi.org/10.3389/fmicb.2021.793037

90. Liu, H., Zhang, X., Zhang, H., Yao, X., Zhou, M., Wang, J., et al. (2018). Effect of air pollution on the total bacteria and pathogenic bacteria in different sizes of particulate matter. *Environmental Pollution, 233*, 483–493. https://doi.org/10.1016/j.envpol.2017.10.070
91. Kim, T. Y., Kim, J. M., Lee, H. L., Go, M. J., Joo, S. G., Kim, J. H., et al. (2023). Codium fragile suppresses PM(2.5)-induced cognitive dysfunction by regulating gut-brain axis via TLR-4/MyD88 pathway. *International Journal of Molecular Sciences, 24*(16), 12898. https://doi.org/10.3390/ijms241612898
92. Margaryan, A., Panosyan, H., & Birkeland, N.-K. (2021). Heavy metal resistance in prokaryotes: Mechanism and application. In D. Egamberdieva et al. (Eds.), *Microbial communities and their interactions in the extreme environment* (pp. 273–313). Springer Singapore.
93. Odenwald, M. A., & Turner, J. R. (2017). The intestinal epithelial barrier: A therapeutic target? *Nature Reviews. Gastroenterology & Hepatology, 14*(1), 9–21. https://doi.org/10.1038/nrgastro.2016.169
94. Panda, S. R., Chaudhari, V. B., Ahmed, S., Kwatra, M., Jala, A., Ponneganti, S., et al. (2023). Ambient particulate matter (PM(2.5)) exposure contributes to neurodegeneration through the microbiome-gut-brain axis: Therapeutic role of melatonin. *Environmental Toxicology and Pharmacology, 101*, 104183. https://doi.org/10.1016/j.etap.2023.104183
95. Dai, S., Wang, Z., Yang, Y., Du, P., & Li, X. (2022). PM(2.5) induced weight loss of mice through altering the intestinal microenvironment: Mucus barrier, gut microbiota, and metabolic profiling. *Journal of Hazardous Materials, 431*, 128653. https://doi.org/10.1016/j.jhazmat.2022.128653
96. Ran, Z., An, Y., Zhou, J., Yang, J., Zhang, Y., Yang, J., et al. (2021). Subchronic exposure to concentrated ambient PM2.5 perturbs gut and lung microbiota as well as metabolic profiles in mice. *Environmental Pollution, 272*, 115987. https://doi.org/10.1016/j.envpol.2020.115987
97. Byrne, C. S., Chambers, E. S., Morrison, D. J., & Frost, G. (2015). The role of short chain fatty acids in appetite regulation and energy homeostasis. *International Journal of Obesity, 39*(9), 1331–1338. https://doi.org/10.1038/ijo.2015.84
98. Liu, H., Wang, G., Zhang, J., Lu, B., Li, D., & Chen, J. (2024). Inhalation of diesel exhaust particulate matter induces locomotor hyperactivity and its relationship with brain and gut metabolism. *Environment International, 183*, 108359. https://doi.org/10.1016/j.envint.2023.108359
99. Park, S. K., Kang, J. Y., Kim, J. M., Kim, M. J., Lee, H. L., Moon, J. H., et al. (2022). Water extract of ecklonia cava protects against fine dust (PM(2.5))-induced health damage by regulating gut health. *Journal of Microbiology and Biotechnology, 32*(7), 927–937. https://doi.org/10.4014/jmb.2203.03020
100. Thomas, R. H., Meeking, M. M., Mepham, J. R., Tichenoff, L., Possmayer, F., Liu, S., & MacFabe, D. F. (2012). The enteric bacterial metabolite propionic acid alters brain and plasma phospholipid molecular species: Further development of a rodent model of autism spectrum disorders. *Journal of Neuroinflammation, 9*, 153. https://doi.org/10.1186/1742-2094-9-153
101. Liu, F., Li, J., Wu, F., Zheng, H., Peng, Q., & Zhou, H. (2019). Altered composition and function of intestinal microbiota in autism spectrum disorders: A systematic review. *Translational Psychiatry, 9*(1), 43. https://doi.org/10.1038/s41398-019-0389-6
102. MacFabe, D. F. (2015). Enteric short-chain fatty acids: Microbial messengers of metabolism, mitochondria, and mind: Implications in autism spectrum disorders. *Microbial Ecology in Health and Disease, 26*, 28177. https://doi.org/10.3402/mehd.v26.28177
103. Al-Lahham, S. H., Peppelenbosch, M. P., Roelofsen, H., Vonk, R. J., & Venema, K. (2010). Biological effects of propionic acid in humans; metabolism, potential applications and underlying mechanisms. *Biochimica et Biophysica Acta, 1801*(11), 1175–1183. https://doi.org/10.1016/j.bbalip.2010.07.007
104. Ristori, M. V., Quagliariello, A., Reddel, S., Ianiro, G., Vicari, S., Gasbarrini, A., & Putignani, L. (2019). Autism, gastrointestinal symptoms and modulation of gut microbiota by nutritional interventions. *Nutrients, 11*(11), 2812. https://doi.org/10.3390/nu11112812

105. Qin, L., Wu, X., Block, M. L., Liu, Y., Breese, G. R., Hong, J. S., et al. (2007). Systemic LPS causes chronic neuroinflammation and progressive neurodegeneration. *Glia, 55*(5), 453–462. https://doi.org/10.1002/glia.20467
106. Feil, R., & Fraga, M. F. (2012). Epigenetics and the environment: Emerging patterns and implications. *Nature Reviews. Genetics, 13*(2), 97–109. https://doi.org/10.1038/nrg3142
107. Brown, J. M., Terada, L. S., Grosso, M. A., Whitman, G. J., Velasco, S. E., Patt, A., et al. (1988). Hydrogen peroxide mediates reperfusion injury in the isolated rat heart. *Molecular and Cellular Biochemistry, 84*(2), 173–175. https://doi.org/10.1007/BF00421052
108. Mukherjee, S., Dasgupta, S., Mishra, P. K., & Chaudhury, K. (2021). Air pollution-induced epigenetic changes: Disease development and a possible link with hypersensitivity pneumonitis. *Environmental Science and Pollution Research International, 28*(40), 55981–56002. https://doi.org/10.1007/s11356-021-16056-x
109. Alfano, R., Herceg, Z., Nawrot, T. S., Chadeau-Hyam, M., Ghantous, A., & Plusquin, M. (2018). The impact of air pollution on our epigenome: How far is the evidence? (A systematic review). *Current Environmental Health Reports, 5*(4), 544–578. https://doi.org/10.1007/s40572-018-0218-8
110. Li, Z., Liang, D., Ebelt, S., Gearing, M., Kobor, M. S., Konwar, C., et al. (2024). Differential DNA methylation in the brain as potential mediator of the association between traffic-related PM(2.5) and neuropathology markers of Alzheimer's disease. *Alzheimers Dement, 20*(4), 2538–2551. https://doi.org/10.1002/alz.13650
111. Zhou, Y., Zhang, M., Liu, W., Li, Y., Qin, Y., & Xu, Y. (2020). Transgenerational transmission of neurodevelopmental disorders induced by maternal exposure to PM2.5. *Chemosphere, 255*, 126920. https://doi.org/10.1016/j.chemosphere.2020.126920
112. Rider, C. F., & Carlsten, C. (2019). Air pollution and DNA methylation: Effects of exposure in humans. *Clinical Epigenetics, 11*(1), 131. https://doi.org/10.1186/s13148-019-0713-2
113. Liu, X. S., Wu, H., Ji, X., Stelzer, Y., Wu, X., Czauderna, S., et al. (2016). Editing DNA methylation in the mammalian genome. *Cell, 167*(1), 233–247.e17. https://doi.org/10.1016/j.cell.2016.08.056
114. Polli, A., Ghosh, M., Bakusic, J., Ickmans, K., Monteyne, D., Velkeniers, B., et al. (2020). DNA methylation and brain-derived neurotrophic factor expression account for symptoms and widespread hyperalgesia in patients with chronic fatigue syndrome and comorbid fibromyalgia. *Arthritis & Rhematology, 72*(11), 1936–1944. https://doi.org/10.1002/art.41405
115. Ng, T. K. S., Ho, C. S. H., Tam, W. W. S., Kua, E. H., & Ho, R. C. (2019). Decreased serum brain-derived neurotrophic factor (BDNF) levels in patients with Alzheimer's disease (AD): A systematic review and meta-analysis. *International Journal of Molecular Sciences, 20*(2), 257. https://doi.org/10.3390/ijms20020257
116. Franzmeier, N., Ren, J., Damm, A., Monte-Rubio, G., Boada, M., Ruiz, A., et al. (2021). The BDNF(Val66Met) SNP modulates the association between beta-amyloid and hippocampal disconnection in Alzheimer's disease. *Molecular Psychiatry, 26*(2), 614–628. https://doi.org/10.1038/s41380-019-0404-6
117. Greer, E. L., & Shi, Y. (2012). Histone methylation: A dynamic mark in health, disease and inheritance. *Nature Reviews. Genetics, 13*(5), 343–357. https://doi.org/10.1038/nrg3173
118. Heintzman, N. D., Stuart, R. K., Hon, G., Fu, Y., Ching, C. W., Hawkins, R. D., et al. (2007). Distinct and predictive chromatin signatures of transcriptional promoters and enhancers in the human genome. *Nature Genetics, 39*(3), 311–318. https://doi.org/10.1038/ng1966
119. Zhang, Y., Sun, Z., Jia, J., Du, T., Zhang, N., Tang, Y., et al. (2021). Overview of histone modification. *Advances in Experimental Medicine and Biology, 1283*, 1–16. https://doi.org/10.1007/978-981-15-8104-5_1
120. Calderon-Garciduenas, L., Herrera-Soto, A., Jury, N., Maher, B. A., Gonzalez-Maciel, A., Reynoso-Robles, R., et al. (2020). Reduced repressive epigenetic marks, increased DNA damage and Alzheimer's disease hallmarks in the brain of humans and mice exposed to particulate urban air pollution. *Environmental Research, 183*, 109226. https://doi.org/10.1016/j.envres.2020.109226

121. Li, J., Zhang, L., Wang, J., Jia, R., Zhang, X., Li, X., et al. (2022). Differential expression of long non-coding RNAs in the hippocampus of mice exposed to PM(2.5) in Dalian, China. *Environmental Science and Pollution Research International, 29*(8), 12136–12146. https://doi.org/10.1007/s11356-021-16496-5
122. Wegel, E., & Shaw, P. (2005). Gene activation and deactivation related changes in the three-dimensional structure of chromatin. *Chromosoma, 114*(5), 331–337. https://doi.org/10.1007/s00412-005-0015-7
123. Gonzalez-Maciel, A., Reynoso-Robles, R., Torres-Jardon, R., Mukherjee, P. S., & Calderon-Garciduenas, L. (2017). Combustion-derived nanoparticles in key brain target cells and organelles in young urbanites: Culprit hidden in plain sight in Alzheimer's disease development. *Journal of Alzheimer's Disease, 59*(1), 189–208. https://doi.org/10.3233/JAD-170012
124. Chen, F., Zhang, W., Mfarrej, M. F. B., Saleem, M. H., Khan, K. A., Ma, J., et al. (2024). Breathing in danger: Understanding the multifaceted impact of air pollution on health impacts. *Ecotoxicology and Environmental Safety, 280*, 116532. https://doi.org/10.1016/j.ecoenv.2024.116532
125. Cory-Slechta, D. A., Merrill, A., & Sobolewski, M. (2023). Air pollution-related neurotoxicity across the life span. *Annual Review of Pharmacology and Toxicology, 63*, 143–163. https://doi.org/10.1146/annurev-pharmtox-051921-020812

Chapter 5
Air Pollution and Neurodevelopment

Hui-Xian Zeng, Xiao-Wen Zeng, and Guang-Hui Dong

Keywords Air pollution · Neurodevelopmental disorders · Toxic mechanisms · Sensitive exposure windows · White matter damage

5.1 Introduction

Exposure to air pollution has been associated with adverse health effects on the brain. Ongoing brain development may render children especially susceptible to air pollution. Epidemiological studies increasingly link air pollution to multiple neurodevelopmental disorders (NDDs), human neuroimaging research offers the evidence between alterations in the growing brains of children and air pollution, and animal studies further reveal possible mechanisms of neurotoxic effects of air pollution. Several mechanisms of neurotoxicity induced by air pollution are shared in multiple NDDs. Future studies examining the responsible pollutants of air pollution underlying this neurotoxicity are crucial to fully understand the etiology of NDDs and safeguard brain health for all youth.

5.2 Air Pollution and Neurodevelopment: A Public Health Concern

Air pollution is now the second leading risk factor for death worldwide, contributing to 8.1 million deaths globally in 2021 [24]. The updated WHO Global Air Quality Guidelines (AQGs) recommend air quality levels for six air pollutants,

H.-X. Zeng · X.-W. Zeng · G.-H. Dong (✉)
Joint International Research Laboratory of Environment and Health, Ministry of Education, Guangdong Provincial Engineering Technology Research Center of Environmental Pollution and Health Risk Assessment, Department of Occupational and Environmental Health, School of Public Health, Sun Yat-sen University, Guangzhou, China
e-mail: donggh5@mail.sysu.edu.cn

K. M. Kanninen, A. R. White (eds.), *Air Pollution and the Brain*,
https://doi.org/10.1007/978-3-031-99302-2_5

where evidence advances the most on health effects from exposure. These air pollutants include particulate matter of different size fractions (PM_{10} and $PM_{2.5}$), nitrogen dioxide (NO_2), ozone (O_3), sulfur dioxide (SO_2), and carbon monoxide (CO). Furthermore, WHO's new guidelines consider certain types of PM [e.g., black carbon or elemental carbon (BC/EC), ultrafine particles (UFP), and particles originating from sand and dust storms (SDS)] as harmful air pollutants because of their potential effects on human health, although they lack recommended standards. However, 99% of global people daily breathe air that is not as clean as what the WHO recommends [56]. According to statistics from the WHO in 2018, 93% of global children under 15 years of age breathe polluted air [54]. Pediatric populations are especially susceptible to air pollution due to their immature natural defense systems and higher ratio of breathing rate to body size. There is robust evidence to demonstrate an apparent association between air pollution exposure and the risk to children's health, especially as regards respiratory and cardiovascular diseases. Importantly, current evidences establish connections between air pollution exposure and neurological diseases in children [17]. According to the Global Burden of Disease Study 2021, an estimated 3.40 billion people (about 43.1% of the global population) had a condition affecting the nervous system [14]. Paediatric neurological conditions and NDDs contributed 80.3 million disability-adjusted life-years (DALYs) or 18·2% of the total DALYs attributable to neurological health loss in 2021 [14]. As the prevalence of NDDs has risen alarmingly on a global scale in recent years, it is timely and vital to assess whether air pollution raises developmental risks.

Neurodevelopment is a term referring to the brain development of systems or networks responsible for perceptual, motor, and cognitive functioning, such as learning, memory, executive function, social skills, and overall brain function. Neurodevelopment continues well after birth into early adolescence. Deficits in neurodevelopment at any stage can contribute to NDDs. NDDs are a class of disorders affecting brain development and function, causing difficulties in social, cognitive, and emotional functioning. There are many different types of NDDs, including autism spectrum disorder (ASD), attention-deficit/hyperactive disorder (ADHD), communication disorders, conduct disorders, intellectual disabilities, learning disorders, and neurodevelopmental motor disorders. The complex interactions between genetic and environmental factors are involved in the onset of NDDs. Pregnancy and early childhood are periods with high neuroplasticity in brain development. However, there is evidence suggesting that exposure to pollutants during these sensitive windows can have lifelong neurodevelopmental consequences. Current research exhibits compelling relationships between air pollution exposure and the development of childhood NDDs. These associations are alarming because brain development occurs from a few weeks after conception to the second decade of life, and a significant number of global children routinely are exposed to polluted air.

5.3 Epidemiological Evidence: Air Pollution and Neurodevelopmental Disorders

Extensive studies have been conducted to explore the potential effects of air pollutants, namely different-sized PM and NO_2, on neurodevelopment in children. An increasing corpus of epidemiological research has indicated an association between air pollution and the vulnerability of children to NDDs. In the following section, we have summarized the effects of air pollutants on multiple types of NDDs during different exposure windows.

5.3.1 Air Pollution and Autism Spectrum Disorder

ASD is defined by repetitive stereotyped behaviors and difficulties in social interaction and communication [27]. The term ASD includes autism and a range of similar disorders. Although ASD can be diagnosed at any age, it is described as a "developmental disorder" because symptoms generally manifest during the first 2 years of life. Globally, about 1/100 children were diagnosed with ASD in 2021 [58]. There is a notable sex difference in this prevalence, with boys having a 4–5 times higher prevalence than girls [55]. There are many causes that make a child more likely to develop ASD, including genetic and environmental factors. Environmental factors, such as exposure to air pollution, organophosphates, and polychlorinated biphenyls, contribute to the onset and development of ASD [26].

Air pollution has been a global environmental issue. A growing number of research has shown links between air pollution and an increased risk of ASD. In recent years, evidence from meta-analyses suggested a positive overall association between air PM and childhood ASD [20, 31, 32]. In the most recent meta-analysis, Liu et al. [32] investigated the associations between ambient PM and ASD during pregnancy and infancy by performing a systematic literature search using eight databases. They included 27 studies with 1,009,799 participants to perform a meta-analysis. In this meta-analysis, the estimated effects were combined separately according to different types of PM. They found that per 10 μg/m^3 increase of PM_1 [$N = 1$ study, odds ratio (OR) = 3.64, 95% CI: 1.20, 11.08], $PM_{2.5}$ ($N = 15$ studies, OR = 1.34, 95% CI: 1.12, 1.60), and PM_{coarse} ($N = 3$ studies, OR = 1.06, 95% CI: 1.02, 1.11) was linked to an increased risk of ASD, respectively, while no association with PM_{10} was found ($N = 11$ studies, OR = 1.03, 95% CI: 0.97, 1.09). They also observed that $PM_{2.5}$ exposure increased the risk of ASD in boys ($N = 4$ studies, OR = 1.26, 95% CI: 1.12, 1.42). The authors noted that the combined results of different sizes of PM differed in study design type and study location. Positive associations were found between $PM_{2.5}$ and ASD in both case-control ($N = 10$ studies, OR = 1.43, 95% CI: 1.09, 1.87) and cohort studies ($N = 5$ studies, OR = 1.24, 95% CI: 1.13, 1.35), but only in case-control studies for PM_{coarse} ($N = 2$ studies, OR = 1.06, 95% CI: 1.02, 1.11). In addition, the authors noted significant associations between

PM types and ASD during different exposure windows. Specifically, exposure to $PM_{2.5}$ in the first trimester ($N = 7$ studies, OR = 1.13, 95% CI: 1.06, 1.21), second trimester ($N = 7$ studies, OR = 1.29, 95% CI: 1.03, 1.62), third trimester ($N = 7$ studies, OR = 1.24, 95% CI: 1.05, 1.46), first year after birth ($N = 6$ studies, OR = 1.93, 95% CI: 1.30, 2.87), and after 1 year of life ($N = 3$ studies, OR = 3.48, 95% CI: 1.77, 6.87) were associated with increased risks of childhood ASD. Overall, this meta-analysis supports the hypothesis that $PM_{2.5}$ exposure increases the risk of ASD in children, and that the exposure time from the first, second, and third trimesters to the first year postpartum may be sensitive windows for $PM_{2.5}$ exposure. These findings may offer more thorough and reliable information on the potential links between various PM types and ASD, which may also assist in identifying the sensitive windows of PM exposure.

In addition to PM exposure, there is moderate evidence to show the association between air gaseous pollutants and the vulnerability of children to ASD. For example, a recent meta-analysis included 14 studies (published before March 1, 2024) and explored the effects of early-life exposure to gases [NO_2, nitrogen oxides (NOx), and O_3] on childhood ASD. The findings demonstrated a moderate association between ASD and exposure to NO_2 ($N = 5$ studies, OR = 1.13, 95 CI%, 0.77, 1.55), O_3 ($N = 5$ studies, OR = 0.82, 95 CI%, 0.49, 1.14), and NO_x ($N = 5$ studies, OR = 1.07, 95 CI%, 0.94, 1.21) [5]. Dutheil et al. [20] carried out a systematic review and meta-analyses of the risk of ASD in newborns following air pollution exposure during the perinatal period. They found that per 10 $\mu g/m^3$ increase of NOx during pregnancy increased the risk of ASD ($N = 8$ studies, OR = 1.02, 95% CI, 1.01, 1.03) [20]. The effects of gaseous contaminants on childhood ASD remain unclear and require further investigations. The summary of meta-analyses in research on ASD and air pollution during different exposure windows is provided in Table 5.1.

5.3.2 Air Pollution and Attention Deficit Hyperactivity Disorder

With a 5% prevalence in children globally, ADHD is characterized by persistent symptoms of inattention and/or hyperactivity and impulsivity that interfere with functioning or development normally [15]. ADHD symptoms begin before the age of twelve, and they become apparent as early as three in certain children. ADHD frequently co-occurs with other mental health issues and has a substantial negative impact on the affected individual, their family, and the community [45]. Although the heritability of ADHD is estimated to be between 70% and 80% [22], the combination of other social and environmental determinants is also involved in the onset and development of ADHD [15]. Notably, air pollution may be as an early-life risk factor of childhood ADHD.

Table 5.1 Meta-analyses on risk of ASD for air pollution during different exposure window

Pollutants	Exposure window	Studies	Increment	ORs (95%CI)	References
PM_1	First 3 years of life	1	10 μg/m³	3.64 (1.12, 11.08)	[32]
$PM_{2.5}$	Preconception	3	5 μg/m³	1.17 (1.00, 1.33)	[20]
	Pregnancy	8		1.05 (1.03, 1.07)	
	Postnatal	7		1.11 (1.06, 1.15)	
$PM_{2.5}$	Preconception	5	10 μg/m³	1.01 (0.80, 1.28)	[31]
	First trimester	10		1.07 (1.02, 1.13)	
	Second trimester	10		1.12 (0.93, 1.35)	
	Third trimester	10		1.20 (1.05, 1.37)	
	Pregnancy	13		1.32 (1.03, 1.69)	
	First year after birth	9		1.62 (1.22, 2.15)	
	Second year after birth	3		3.13 (1.47, 6.67)	
$PM_{2.5}$	Preconception	4	10 μg/m³	1.05 (0.83, 1.34)	[32]
	First trimester	7		1.13 (1.06, 1.21)	
	Second trimester	7		1.29 (1.03,1.62)	
	Third trimester	7		1.24 (1.05, 1.46)	
	Pregnancy	12		1.21 (0.95, 1.56)	
	First year after birth	6		1.93 (1.30, 2.87)	
	After 1 year of life	3		3.48 (1.77, 6.87)	
PM_{10}	Preconception	2	10 μg/m³	1.01 (0.93, 1.09)	[32]
	First trimester	5		0.99 (0.94, 1.05)	
	Second trimester	5		1.03 (0.95, 1.11)	
	Third trimester	5		1.07 (0.97, 1.18)	
	Pregnancy	9		1.01 (0.94,1.08)	
	First year after birth	5		1.06 (0.93, 1.19)	
	After 1 year of life	1		2.81 (1.45, 5.46)	
PM_{coarse}	Preconception	1	10 μg/m³	1.03 (0.94,1.13)	[32]
	Pregnancy	3		1.10 (1.01,1.19)	
	Postnatal	1		1.05 (0.99,1.13)	
NOx	Pregnancy	8	10 μg/m³	1.02 (1.01, 1.03)	[20]
	Postnatal	4		1.02 (0.98, 1.06)	

Abbreviation: PM_1 Particulate matter with aerodynamic diameter ≤ 1 μm, $PM_{2.5}$ Particulate matter with aerodynamic diameter ≤ 2.5 μm, PM_{10} Particulate matter with aerodynamic diameter ≤ 10 μm, PM_{coarse} Particulate matter with aerodynamic diameter between 2.5 and 10 μm, *NOx* Nitrogen oxide

Several systematic reviews offer limited evidence of an association between exposure to air pollution and childhood ADHD [1, 50, 59]. Among these systematic reviews, Aghaei et al. included the most published articles (28 studies published up to April 2018) and the most air pollutants [ambient gaseous (NO_2 and SO_2) and PM (PM_{10}, $PM_{2.5}$, PM_7, polycyclic aromatic hydrocarbons (PAHs), BC/EC)]. They did not perform meta-analyses dut to variations and differences in methodology used in studies (e.g., time windows of exposure, exposure assessment methods, and outcome measurement). This systematic review found a significant association between particulate air pollutants and increased risk of ADHD (6 of 8 studies on PM, 3 of 6 studies on EC/BC, 1 of 7 studies on PAHs) as well as an association between childhood ADHD and gaseous air pollution (6 of 12 studies on NO_2, 2 of 2 studies on SO_2). These findings emphasize the importance of future research to use standard methods for exposure and outcome assessment.

The birth cohort studies conducted in various regions provide clear associations between $PM_{2.5}$, PM_{10}, and NO_2 with childhood ADHD. For example, a birth cohort study with 10-year follow up (2003–2012) conducted in Korea (8936 infants and 314 cases) reported that exposure to air pollution from birth increased the risk of childhood ADHD [hazard ratio (HR) of per 1 μg/m^3 increase of pollutants: PM_{10} = 1.18 (95% CI: 1.15, 1.21) and NO_2 = 1.03 (95% CI: 1.02, 1.04)] [37]. Another birth cohort study conducted in Saxony (66,823 infants born in 2000–2004 and followed until 2014, and 2044 cases) showed similar associations between ADHD and per 10 μg/m^3 increase of PM_{10} (HR = 1.97, 95% CI: 1.25, 2.86) or NO_2 (HR = 1.32, 95% CI: 1.10, 1.58) [36]. A Taiwan birth cohort (425,736 infants born in 2004–2015, and 9294 cases) exhibited a strong association between ADHD with a 10 μg/m^3 increment of $PM_{2.5}$ in the first trimester during pregnancy (HR = 1.27, 95% CI: 1.14, 1.43) and in the first year after birth (HR = 1.40, 95% CI: 1.40, 1.87) [13]. Large birth cohort studies conducted in Canada (37,000 participants and 1217 cases) and Denmark (809,654 participants and 19,045 cases) also reported positive associations between early life exposure to $PM_{2.5}$ and NO_2 with an increased incidence of ADHD [52, 57]. Overall, Table 5.2 lists the summary of birth cohort studies on air pollution and the incidence of ADHD.

5.3.3 *Air Pollution and Neurodevelopmental Skills*

Neurodevelopmental skills include memory and learning, attention and executive functions, motor and sensorimotor functions, verbal language, intellective functioning, and numeric ability. The consolidation of major neurodevelopmental skills typically begins in preschool. An increasing amount of research reported that exposure to air pollution during fetal life and early life may have detrimental impacts on neurodevelopmental skills in typically developing children. For example, Castagna et al. aimed to evaluate the association between early life exposure to air pollutants and specific neurodevelopmental skills in preschool- and school-children between the ages of 3 and 12. They performed a systematic literature search via databases until September 2021, and included 30 studies for a systematic review [12]. This systematic review observed that the most affected domains in children after early life exposure to air pollutants were the global intellective functioning (prenatal exposure: 3 of 3 studies on PAHs, 1 of 2 studies on $PM_{2.5}$, 1 of 1 study on NO_2; postnatal exposure: 1 of 1 study on $PM_{2.5}$, 1 of 1 study on BC) and attention/executive functions (prenatal exposure: 1 of 1 study on PAHs, 2 of 2 studies on $PM_{2.5}$, 1 of 1 study on BC; postnatal exposure: 1 of 2 studies on PAHs, 2 of 2 studies on $PM_{2.5}$, 1 of 2 studies on PM_{10}, 1 of 5 studies on NO_2, 2 of 3 studies on BC). Shang et al. performed a meta-analysis to explore the association between prenatal NO_2

Table 5.2 Summary of birth cohort studies on air pollution and the incidence of ADHD

Pollutants	Region	Exposure window	Number	HRs (95% CI)	References
PM_{10}, NO_2	Korea (nationwide)	A 10-year follow-up from birth (individuals born in 2002)	8936 individuals, 314 cases	PM_{10}: 1.18 (1.15, 1.21)[a] NO_2: 1.03 (1.02, 1.04)[a]	[37]
PM_{10}, NO_2	Saxony, Germany	A 10- to 14-year follow-up from birth (individuals born in 2000–2004 and followed until 2014)	66,823 individuals, 2044 cases	PM_{10}: 1.97 (1.25, 2.86)[b] NO_2: 1.32 (1.10, 1.58)[b]	[36]
$PM_{2.5}$	Taiwan, China	The first, second, and third trimesters and from age 1–5 years after birth (individuals born in 2004–2015)	425,736 individuals, 9294 cases	1.27 (1.14, 1.43) for the first trimester[b]; 1.40 (1.40, 1.87) for the first year after birth[b]; 1.87 (1.53, 2.29) for the third year after birth[b]	[13]
$PM_{2.5}$, NO_2	Denmark (nationwide)	The first 5 years of life (individuals born in 1992–2007)	809,654 individuals, 19,045 cases	$PM_{2.5}$: 1.51 (1.41, 1.62)[c] NO_2: 1.38 (1.35, 1.42)[b]	[52]
$PM_{2.5}$, NO_2	Vancouver, Canada	A 7-year follow-up from birth (individuals born in 2000–2001)	37,000 individuals, 1217 cases	$PM_{2.5}$:1.11 (1.06, 1.17)[d] NO_2: 1.01 (0.96, 1.07)[d]	[57]

Note: [a]HR (95% CI) per 1 μg/m^3 increment. [b]HR (95% CI) per 10 μg/m^3 increment. [c]HR (95% CI) per 5 μg/m^3 increment. [d]HR (95% CI) per interquartile range increment

exposure and childhood neurodevelopment. This meta-analysis included 10 studies published until May 2019. The findings reported that per 10 μg/m^3 increment of NO_2 during pregnancy was linked to a −0.76 (95% CI, −1.34, −0.18) point decline in global psychomotor and a −0.62 (95% CI, −1.09, −0.16) point decline in fine psychomotor for children, and that prenatal NO_2 exposure was not linked to general cognitive and language [49]. Overall, these findings indicate that early life exposure to air pollution, especially PAHs, $PM_{2.5}$, BC, and NO_2, is associated with difficulties in later neurodevelopmental skills. These findings further support the need for global action to reduce air pollution exposure, especially during pregnancy and the early stages of child development.

5.4 Neuroimaging Evidence: Air Pollution and Neurodevelopment

Neuroimaging studies offer the chance to examine the human brain in vivo across the lifespan, and magnetic resonance imaging (MRI) is utilized frequently to investigate the neurodevelopmental effects of exposure to ambient air pollution. Most human neuroimaging studies estimate air pollution exposure levels for the child based on their residential location using a high-resolution spatiotemporal modeling. The findings of these neuroimaging studies not only support the effects of air pollution on the developing brain but also show evidence of brain changes that is consistent with known cerebral alterations in multiple NDDs. Overall, there is compelling evidence linking exposure to air pollutants (e.g., $PM_{2.5}$ and its components, O_3, and NO_x) and various brain differences in children, including brain macrostructure, gray and white matter microstructure, brain metabolites, brain connectivity and network, and cerebral blood flow. Table 5.3 lists the summary of neuroimaging studies on air pollution and neurodevelopment.

5.4.1 Brain Structure

MRI research offers the opportunity to assess structural alterations in the brains of children after exposure to air pollution. These studies indicate that exposure to air pollution induces neuroanatomically localized changes, not global differences in brain structure. For example, Peterson et al. studied the effects of prenatal exposure to air pollution on brain structure in 332 US African-American or Dominican youth (6–14 years old). They found that both exposure to $PM_{2.5}$ and PAHs was associated with thinner dorsal parietal cortices, thickener postero–inferior and mesial wall cortices, smaller white matter volumes, lower organization in white matter of the internal capsule and frontal lobe, greater microstructural organization in subcortical gray matter nuclei, and lower cortical blood flow in children. They also noted that associations were stronger for $PM_{2.5}$ in boys and PAHs in girls, and that cortico-striato-thalamocortical circuits were comprised of these affected regions [43]. The cortico-striato-thalamocortical circuits can be involved in the regulation of thinking, emotion, and behavior. In addition, several studies also reported that exposure to air pollution [e.g., PM_{10}, PM_{coarse}, $PM_{2.5}$ and its components (EC, OC, and PAHs), and NO_2] was associated with differences in the volume of cortical white matter, cortical grey matter, cerebellum, hippocampus, amygdala, corpus callosum, caudate nucleus, as well as differences in hemispheric specific alterations in gray matter of the frontal, parietal, temporal and occipital lobes (Table 5.3). The human frontal, parietal, and temporal areas continue to develop well into late adolescence and early adulthood, supporting the maturation of higher-order executive functions. Hence, brain regional effects of exposure to air pollution may shed light on downstream behavioral consequences, such as poor neurodevelopmental skills.

5.4.2 Differences in White Matter Tracts

Diffusion-weighted imaging (DWI) is a valuable tool for assessing the integrity and ultrastructural properties of brain white matter, and has great promise in identifying early alterations in a variety of neurological disorders. Diffusion-weighted imaging

Table 5.3 Summary of neuroimaging studies on air pollution in children

Pollutant	Exposure window	Number	Major findings	References
Air pollutants	Pregnancy	3133 children	PM_{coarse}: larger cerebellum; PAHs: smaller hippocampus and amygdala; OC: smaller amygdala.	[34]
Air pollutants	Pregnancy	783 children	$PM_{2.5}$: thinner cortical thickness; PM_{coarse}: thinner cortical thickness.	[23]
Air pollutants	Pregnancy	1186 children	PAHs: larger cortical white matter in APOEε4 carriers; PM_{coarse}: larger cortical white matter in APOEε4 carriers or in children with a high PRS for AD.	[21]
$PM_{2.5}$ and PAHs	Pregnancy	332 children	Smaller white matter volumes; Thickener postero-inferior and mesial wall cortices; Thinner dorsal parietal cortices; Higher cortical N-acetyl-L-aspartate; Lower cortical blood flow; Greater microstructural organization in subcortical gray matter nuclei.	[43]
PM_{10}, $PM_{2.5}$, and NO_2	Pregnancy	469 infants	PM_{10}: smaller cortical grey matter, hippocampus, and amygdala; larger cerebellum.	[8]
$PM_{2.5}$	Pregnancy	186 children	Smaller corpus callosum.	[38]
$PM_{2.5}$	Childhood	10,343 children	Hemispheric specific differences in gray matter of the frontal, parietal, temporal and occipital lobes as well as subcortical and cerebellum brain regions.	[19]
PAHs	Pregnancy	40 children	Reduced white matter surface in the left hemisphere.	[44]
PAHs and BPA	Childhood	242 children	Smaller caudate nucleus volume.	[39]
TRAP	Childhood	163 children	PAHs, EC, and NO_2: smaller caudate volume in *APOE*ε4 allele carriers.	[2]
ECAT	Childhood	135 children	Thinner cortical thickness; Smaller gray matter volume in the left pre- and post-central gyri, the cerebellum, and inferior parietal lobe.	[6]
Air pollutants	Pregnancy	2954 children	Lower global FA: NO_X, PM_{10}, $PM_{2.5}$; Higher global MD: NO_2, PM_{10}, $PM_{2.5}$, UFP, Cu, Fe, Si.	[33]
	Childhood		Lower global FA: NO_2, NO_X, Cu, K; Higher global MD: NO_2, NO_X, PM_{10}, $PM_{2.5}$, K, Si, Zn.	
$PM_{2.5}$ and NO_2	From pregnancy to preadolescence	3515 children	$PM_{2.5}$: larger putamen volume, lower global FA, and higher global MD; NO_2: lower global FA, higher global MD.	[7]

(continued)

Table 5.3 (continued)

Pollutant	Exposure window	Number	Major findings	References
$PM_{2.5}$	Childhood	7602 children	Increased restricted isotropic diffusion in the left cingulum, in the left superior longitudinal fasciculus, and bilaterally in the fornix and uncinate fasciculus; Decreased MD of white matter tracts.	[10]
$PM_{2.5}$, O_3, and NO_2	Childhood	8796 children	$PM_{2.5}$: greater isotropic intracellular diffusion in the thalamus, brainstem, and accumbens.	[51]
ECAT	Childhood	145 children	Increased myo-inositol in the anterior cingulate cortex.	[9]
$PM_{2.5}$	Childhood	57 children	Low hippocampal NAA/Cr ratio.	[11]
EC and NO_2	Childhood	263 children	Lower functional integration and segregation in key brain networks relevant to both inner mental processes and stimulus-driven mental operations.	[47]
TRAP	From pregnancy to preadolescence	2197 children	NO_2, NO_x, and $PM_{2.5}$ absorbance: higher correlation coefficients between brain areas.	[42]
$PM_{2.5}$, O_3, and NO_2	Childhood	9497 children	$PM_{2.5}$: greater inter-network and subcortical-to-network FC; O_3: greater intra-network and less subcortical-to-network FC; NO_2: less inter-network and subcortical-to-network FC.	[18]
Cu	Childhood	263 children	A reciprocal connectivity reduction between the caudate nucleus and the frontal cortex; Altered structure of the basal ganglia; Higher FA in white matter close to the caudate nucleus.	[46]

Abbreviation: $PM_{2.5}$ Particulate matter with aerodynamic diameter ≤ 2.5 μm, PM_{coarse} Particulate matter with aerodynamic diameter between 2.5 and 10 μm, PM_{10} Particulate matter with aerodynamic diameter ≤ 10 μm, *FC* Functional connectivity, *FA* Fractional anisotropy, *MD* Mean diffusivity, *TRAP* Traffic-related air pollution, *PRS* Polygenic risk score, *AD* Alzheimer's disease, *ECAT* Elemental carbon attributable to traffic, *EC* element carbon, *OC* Organic carbon, *PAHs* Polycyclic aromatic hydrocarbon, *UFP* Ultrafine particles, *Cu* Copper, *Fe* Iron, *K* potassium, *Si* Silicon, NO_2 Nitrogen dioxide, *NOx* Nitrogen oxide, *NAA/Cr* N-acetylaspartate/creatine

includes diffusion tensor imaging (DTI) and restricted spectrum imaging (RSI). Studies of DTI reveal that exposure to air pollution is linked to lower fractional anisotropy and higher mean diffusivity in the brains of children [7, 33]. For example, a study using DTI aimed to explore whether pregnancy and childhood exposures to air pollution was associated with changes in white matter microstructure in preadolescents at 9–12 years of age (2954 children). This study found that exposure to air pollutants (e.g, PM_{10}, $PM_{2.5}$, NO_2, and NO_X) in pregnancy and childhood was linked to lower fractional anisotropy and higher mean diffusivity in the white matter of preadolescents [33]. Fractional anisotropy quantifies the overall magnitude of water directional movement along the axonal fibers, and mean diffusivity is a measure of the overall diffusion of water molecules in the brain. When damage occurs in white matter tracts, mean diffusivity increases and fractional anisotropy decreases. Damage to white matter indicates less myelin, a lower fiber density, and less directional fiber coherence. In some cases, fractional anisotropy may increase when the white matter is damaged. Pujol et al. used DTI to assess the association between copper of $PM_{2.5}$ in school environments and brain development in 2836 children aged from 8 to 12 years. They reported that exposure to copper of $PM_{2.5}$ was linked to poorer motor performance and altered structure of the basal ganglia as well as higher fractional anisotropy in white matter close to the caudate nucleus [46]. The basal ganglia can play a crucial role in the dopaminergic system, motor planning, and reward processing. In complex structures with tracts crossing in different directions, higher fractional anisotropy may denote less mature or less structured tissue. Another two studies using RSI also found the associations between childhood $PM_{2.5}$ exposure and alterations in white matter microstructure [10, 51]. Burnor et al. assessed associations between annual residential $PM_{2.5}$ exposure and white matter microstructure health in 7602 children between 9 and 10 years of age. They noted that childhood $PM_{2.5}$ exposure was associated with higher restricted isotropic diffusion in the left cingulum, in the left superior longitudinal fasciculus, and bilaterally in the fornix and uncinate fasciculus, as well as lower mean diffusivity of white matter tracts [10]. Overall, these results provide more evidence for a connection between air pollution and white matter damage during brain development.

5.4.3 Brain Metabolites

The changes of brain metabolites can affect neurodevelopment. Studies of proton magnetic resonance spectroscopy (MRS) show the association between brain metabolites and air pollution exposure. For example, a cohort study tested whether hippocampal metabolite ratios were related to air pollution in 57 children. This study observed that the hippocampal N-acetylaspartate/creatine (NAA/Cr) ratio was lower in youth living high-polluted region than in children living in a low-polluted region, and that this decrease was associated with behavioral or cognitive impairment [11]. The NAA/Cr ratio is involved in neuronal density, functional integrity, synaptic function, and cognitive function. Hence, a low NAA/Cr ratio may constitute a spectral marker of cognitive impairment in youths. Another report also revealed that childhood exposure to $PM_{2.5}$ and PAHs increased concentrations of cortical N-acetyl-L-aspartate in children [43]. N-acetyl-L-aspartate can be a marker of neuronal health. In addition, Brunst et al. explored the effects of traffic-related air pollution (TRAP) on brain metabolites in the anterior cingulate cortex in 145 adolescents, and used the fraction of EC attributable to traffic (ECAT, a marker of traffic-related PM) to assess the association with brain metabolites. They found that

ECAT exposure was linked to significant increases in myo-inositol and that both ECAT exposure and elevated myo-inositol levels were related to increased generalized anxiety symptoms. They further noted that myo-inositol levels mediated 12% of the total effect between ECAT and generalized anxiety symptoms [9]. Myoinositol is a metabolite found in glial cells and a component of myelin. It plays a role in the regulation of osmotic pressure and cellular homeostasis. Therefore, air pollution may elicit neuronal dysfunctions and glial inflammatory responses, leading to dysregulated metabolite levels and subsequent behavioral and cognitive impairment.

5.4.4 Brain Connectivity and Network

Recent studies of functional MRI link intrinsic brain connectivity and network to air pollution in childhood. Cotter et al. studied the effect of childhood exposure to air pollutants ($PM_{2.5}$, O_3, and NO_2) on functional brain connectivity in children between the ages of 9 and 10 ($N = 9497$) over a 2-year follow-up period. They found that: (1) higher $PM_{2.5}$ exposure led to greater inter-network and subcortical-to-network functional connectivity; (2) higher O_3 exposure caused greater intra-network, but less subcortical-to-network functional connectivity over time; and (3) higher NO_2 exposure induced less inter-network and subcortical-to-network functional connectivity [18]. Another study explored the associations between TRAP exposure from pregnancy to preadolescence and functional brain connectivity of resting-state networks in preadolescents between the ages of 9 and 12 ($N = 2197$). According to this study, there were stronger correlation coefficients observed in multiple brain areas in preadolescents exposed to higher levels of NO_2, NOx, and $PM_{2.5}$ absorbance from birth to 3 years of age and from 3 to 6 years of age, while the associations were in brain areas predominantly located in the task positive and task negative networks [42]. In addition, Pujol et al. assessed the potential effects of urban pollution during childhood on brain maturation in children from 8 to 12 years ($N = 263$). They noted that higher content of EC and NO_2 was linked to lower functional integration and segregation in key brain networks relevant to both inner mental processes and stimulus-driven mental operations [47]. These findings reveal that exposure to air pollution in childhood appears to adversely affect brain functions at a critical age.

5.4.5 Cerebral Blood Flow

Cerebral blood flow is regulated by neurons and astrocytes and is involved in brain development during critical periods. Studies using arterial spin labeling (ASL) reveal the association between air pollution and cerebral blood flow. For example, Peterson et al. studied the effects of prenatal exposure to air pollution on cerebral blood flow in 332 youth (6–14 years old). They noted that prenatal $PM_{2.5}$ exposure was associated with lower cerebral blood flow in frontal gray matter, lingual and fusiform gyri, hippocampus, and dorsal parietal cortex, with stronger associations in girls than in boys. In addition, they observed that prenatal PAHs exposure was linked to lower cerebral blood flow in the frontal cortex, midcingulate cortex, thalamus, putamen, as well as in the white matter of internal and external capsules, corpus callosum genu, and anterior corona radiata [43]. Given that cellular energy expenditure and cerebral blood flow are closely related, these data suggest that exposure to air pollution, specifically exposure to $PM_{2.5}$ and PAHs, may reduce the resting energy expenditure of neurons or glia within the gray and white matter portions of the brain.

5.5 Animal Experiments: Air Pollution and Neurodevelopment

Animal studies of inhalation exposure to air pollution during key periods for brain development support the results of epidemiological and neuroimaging studies and further provide insights into the mechanisms underlying the relationship between air pollution and neurodevelopment. Several animal studies reported that maternal exposure to air pollution during pregnancy induced neuropathological alterations in the offspring. For example, Klocke et al. studied the neuropathological effects of gestational exposure to concentrated ambient $PM_{2.5}$ and UFP during gestational day 0.5–16.5 (equivalent to human first and second trimesters) on neuropathological consequences in the offspring of B6C3F1 hybrid mice [28, 29]. They found that gestational exposure induced alterations in brain regions (e.g., ventriculomegaly, larger corpus callosum area, and smaller hippocampal volume), corpus callosum hypermyelination, and microglial activation in all offspring at PND 11–15. The study also observed iron deposition in the corpus callosum of female offspring [29]. Moreover, they noted that gestational exposure produced premature and excessive myelination in both offspring sexes at postnatal day (PND) 11–15, and that these changes were consistent with a shift toward a more mature oligodendrocyte phenotype and an active state of microglia in the corpus callosum [28]. They also discovered that gestational exposure induced hypermyelination in the corpus callosum in all offspring as well as ventriculomegaly and high proportion of proliferative oligodendrocyte precursor cells in female offspring at PND 57–61 [28]. Ventriculomegaly is a neuropathological feature of white matter injury, and is related to poor neurodevelopmental outcomes, such as ASD and ADHD. These observations extend our understanding of potential mechanisms by which air pollution exposure may cause white matter damage. Overall, these findings demonstrate that early life exposure to air pollutants induces long-lasting neuropathological deficits in a sex-dependent manner and may be an underlying cause of NDDs.

Early postnatal exposure to air pollution induces poor neurodevelopmental outcomes in rodents. Allen et al. studied the effects of postnatal exposure to UFP during PND 4–7 and PND 10–13 (human third trimester equivalent) on neurodevelopment in C57BL6/J mice [3, 4]. They observed that at PND 14 and PND 55, UFP-exposed male mice generally displayed declines in cytokines (e.g., IL-1β, IL-6, TNFα) in different brain regions, while UFP-exposed female mice showed increases in cytokines in various brain regions. They also noticed higher hippocampal glutamate levels in UFP-exposed male mice, as well as alterations in brain neurotransmitters and glial activation across various brain areas in a sex-dependent manner [3]. In addition, they found that postnatal exposures of mice to UFP induced ventriculomegaly, corpus callosum hypomyelination, smaller corpus callosum size, aberrant white matter development, elevated glutamate levels, excitatory/inhibitory imbalance, microglial activation, amygdala astrocytic activation, and ASD-like repetitive and impulsive behaviors at PND14, with more distinct effects in male mice [4]. These results imply that exposure to air pollution during periods of rapid neurogenesis and gliogenesis may be a risk factor for adverse neurodevelopmental outcomes, especially ASD. Another study explored the impacts of postnatal exposure to $PM_{2.5}$ during PND 8–22 on ASD-like symptoms in male Sprague-Dawley rats. This research reported that postnatal $PM_{2.5}$ exposure induced ASD-like phenotypes (e.g., communication deficits, poor social interaction, and novelty avoidance), increased the releases of proinflammatory cytokines in the hippocampus and prefrontal cortex, promoted astrocytic and microglial activation, and changed ASD candidate gene expression in rat brain at PND 22 [30]. Altogether, these findings show that postnatal exposure to air pollution may induce abnormal behavior, ventriculomegaly, glial activation, neurochemical disruption, and aberrant white matter development in rodents. Furthermore, these findings provide experimental evidence supporting the hypothesis that an etiological association exists between air pollution exposure and the pathophysiology of NDDs.

A series of animal studies shows compelling evidence linking poor neurodevelopment to air pollution exposure from gestation to development. Nephew et al. studied the impacts of exposure to traffic-related PM during gestation (3 weeks) and lactation (3 weeks) on behaviors, cognition, inflammatory markers, and neural integrity in juvenile male Sprague-Dawley rats. Nephew et al. found that traffic-related PM exposure during gestation and lactation led to less social behaviors, more anxiety-related behaviors, cognitive impairment, lower levels of growth and inflammatory factors, and disruption of neural integrity in the juvenile male offspring [40]. Patten et al. exposed Sprague-Dawley rats during gestation and early postnatal development (PND 0–47) to TRAP, and explored NDDs-relevant neuropathological outcomes at PND 51–55 via MRI examination. Patten et al. revealed that direct TARP exposure during gestation and early postnatal development led to sex-dependent differences. These differences included elevated hippocampal neurogenesis in males and more granule cell layer width and smaller lateral ventricle volume in females [41]. Woodward et al. examined neurogenesis of Sprague-Dawley rats exposed to a nano-sized TRAP subfraction from gestation day 2 until 25 weeks of age, and found that TRAP exposure reduced neurogenesis in the dentate gyrus, promoted glial activation, and produced hippocampal iron deposition and microbleeds. In addition, they found that TRAP exposure impaired contextual memory, reduced food-seeking behaviors, and increased depressive behaviors in rats [53]. These findings provide more evidence that exposure to air pollution during gestation and early postnatal development induces behavioral deficits and NDDs-related neuropathological outcomes. Overall, robust evidences from the developmental inhalation toxicology literature show that exposure to air pollution during brain development directly induces poor neurodevelopmental outcomes.

5.6 Possible Mechanisms Underlying Air Pollution-Induced Neurotoxicity

As mentioned above, a growing body of epidemiological, neuroimaging, and animal research shows clear associations between air pollution and NDDs. However, the question of how air pollution may affect so many distinct NDDs raises. Although most research focuses on the distinctions between different NDDs, they share many clinical, neuroimaging, and pathological features, which may be corresponding targets of air pollution. Animal inhalation studies reveal multiple mechanisms of air pollution-induced developmental neurotoxicity [16, 25]. Several mechanisms of developmental neurotoxicity induced by air pollution, including glial activation, inflammation, oxidative stress, neuronal dysfunction, white matter damage, glutamatergic dysfunction, and brain metal dyshomeostasis, are shared in multiple NDDs.

Exposure to air pollution can cause glial activation, which in turn promotes particle deposition in the brain. Microglia, as the main innate immune cells in the central nervous system, are essential for maintaining homeostasis in the brain, identifying changes in the environment, offering neuroprotection, and protecting the brain from harmful substances. Upon the occurrence of an external or internal pathogenic insult in the brain, resting microglia are promptly activated to heal wounded tissues and repair neuronal damage via phagocytosing toxic chemicals and releasing inflammatory mediators. In response to air pollution, microglia undergo phenotypic alterations that lead to amoeboid morphologies, proinflammatory cytokine secretion, and oxidative stress. Air pollution-induced microglial responses also promote the generation of reactive astrocytes. After exposure to air pollution, activated microglia and reactive astrocytes can secrete proinflammatory cytokines and induce the generation of reactive oxygen species, thereby leading to inflammation, oxidative stress, and neuronal dysfunction, which can contribute to poor neurodevelopmental outcomes.

To date, several human MRI studies have found the associations between white matter damage (e.g., alterations in volume and feature) and air pollution in children. Meanwhile, animal inhalation studies report the effects of air pollution on white matter pathology (e.g., ventriculomegaly, abnormal myelination, and changes in oligodendrocyte maturation) [28, 29]. Ventriculomegaly is a clinical feature of periventricular leukomalacia and is indicative of white matter injury resulting to brain interhemispheric dysconnectivity. Oligodendrocytes are essential for the myelination of neuronal axons. Inflammatory responses in microglia and astrocytes can impair the function of oligodendrocytes, which can disrupt myelination in the white matter and further cause white matter damage. White matter is important to coordinate the firing of neuronal assemblies and the synchronization of neural groups within multiple frequencies. Therefore, even slight changes in white matter structure may disrupt cognitive performance during child development. White matter damage can be observed in the brains of children with NDDs. Hence, air pollution-induced white matter damage is involved in the development of NDDs.

Glutamatergic dysfunction is also a prominent target of air pollution. Several studies observe glutamatergic dysfunction in the developing brain with exposure to air pollution [3, 4]. Brain glutamate serves as an immunological mediator, a substrate for metabolism and synthesis, and the primary excitatory neurotransmitter. It also plays a role in various processes, such as cell migration, cell differentiation, cell apoptosis, the regulation of the induction of synapses and their interconnection with astrocytes, and synaptic spatial organization of the cerebellum. However, excessive glutamate is an excitotoxin that can cause lipid peroxidation, white matter damage, and neuronal dysfunction. The sustained rise in brain glutamate levels may indicate the contribution of an excitotoxic mechanism of air pollution.

Air particles contain multiple neurotoxic and trace elements, and particle pollution leads to the depositions of metals in the brain. Certain metals (e.g., iron, copper, and zinc) are necessary for brain function, but other metals (e.g., lead, cadmium, and mercury) are considered non-essential poisons. Recent research on animals and humans suggests that metals (e.g., iron, copper, cadmium) from air particles can impair the blood-brain barrier and then enter the brain [35, 48]. The damage of blood-brain barrier may facilitate metal accumulation in the brain. Both excessive essential metals and non-essential trace metals can exert neurotoxic effects. Maintaining metal homeostasis is important for selective transport across the blood–brain barrier, the cellular communication between astrocytes and neurons, and the elimination of harmful substances via the blood–cerebrospinal fluid barrier. Importantly, air pollution-induced brain metal dyshomeostasis may facilitate a further cascade of downstream consequences such as mitochondrial dysfunction, oxidative stress, and neuronal death, which can promote the onset and development of NDDs.

5.7 Conclusion

Altogether, epidemiological studies provide the association between air pollution and NDDs, human neuroimaging research offers the evidence between variations in the growing brains of children and air pollution, and animal studies reveal possible mechanisms of air pollution-induced neurotoxicity. As of yet, our current understanding of the effects of exposure timing and duration of air pollution on brain development is still limited. Brain maturation is a dynamic process. More research is required to characterize the cumulative impacts of air pollution on brain development during pregnancy, infancy, childhood, and adolescence. Importantly, even at a relatively low level, air pollution still poses a serious risk to growing brains at the early stage of life. In addition, air pollution comprises a complex mixture of gases and particles. The adverse consequences of low-level exposure to air pollution may be determined by pollutant composition rather than by exposure to a single pollutant. The composition of air pollutants can be influenced by the sources of pollution, the interactions between air pollutants, and the distances that air pollutants have traveled. In order to safeguard brain health for all youth, research on both source-specific and mixture effects is required to explore the consequences of air pollution exposure on child development. Today, ambient air pollution is still a global public health issue, especially in low and middle-income countries. We still have a long way to go, and a call for global action is still required to lower ambient air pollution.

References

1. Aghaei, M., Janjani, H., Yousefian, F., et al. (2019). Association between ambient gaseous and particulate air pollutants and attention deficit hyperactivity disorder (ADHD) in children: A systematic review. *Environmental Research, 173*, 135–156.
2. Alemany, S., Vilor-Tejedor, N., García-Esteban, R., et al. (2018). Traffic-related air pollution, APOEε4 status, and neurodevelopmental outcomes among school children enrolled in the BREATHE project (Catalonia, Spain). *Environmental Health Perspectives, 126*, 087001.
3. Allen, J. L., Liu, X., Pelkowski, S., et al. (2014). Early postnatal exposure to ultrafine particulate matter air pollution: Persistent ventriculomegaly, neurochemical disruption, and glial activation preferentially in male mice. *Environmental Health Perspectives, 122*, 939–945.
4. Allen, J. L., Oberdorster, G., Morris-Schaffer, K., et al. (2017). Developmental neurotoxicity of inhaled ambient ultrafine particle air pollution: Parallels with neuropathological and behavioral features of autism and other neurodevelopmental disorders. *Neurotoxicology, 59*, 140–154.
5. Amnuaylojaroen, T., Parasin, N., & Saokaew, S. (2024). Exploring the association between early-life air pollution exposure and autism spectrum disorders in children: A systematic review and meta-analysis. *Reproductive Toxicology, 125*, 108582.
6. Beckwith, T., Cecil, K., Altaye, M., et al. (2020). Reduced gray matter volume and cortical thickness associated with traffic-related air pollution in a longitudinally studied pediatric cohort. *PLoS One, 15*, e0228092.
7. Binter, A. C., Kusters, M. S. W., van den Dries, M. A., et al. (2022). Air pollution, white matter microstructure, and brain volumes: Periods of susceptibility from pregnancy to preadolescence. *Environmental Pollution, 313*, 120109.
8. Bos, B., Barratt, B., Batalle, D., et al. (2023). Prenatal exposure to air pollution is associated with structural changes in the neonatal brain. *Environment International, 174*, 107921.
9. Brunst, K. J., Ryan, P. H., Altaye, M., et al. (2019). Myo-inositol mediates the effects of traffic-related air pollution on generalized anxiety symptoms at age 12 years. *Environmental Research, 175*, 71–78.
10. Burnor, E., Cserbik, D., Cotter, D. L., et al. (2021). Association of outdoor ambient fine particulate matter with intracellular white matter microstructural properties among children. *JAMA Network Open, 4*, e2138300.
11. Calderón-Garcidueñas, L., Mora-Tiscareño, A., Melo-Sánchez, G., et al. (2015). A critical proton MR spectroscopy marker of Alzheimer's disease early neurodegenerative change: Low hippocampal NAA/Cr ratio impacts APOE ε4 Mexico City children and their parents. *Journal of Alzheimer's Disease, 48*, 1065–1075.

12. Castagna, A., Mascheroni, E., Fustinoni, S., et al. (2022). Air pollution and neurodevelopmental skills in preschool- and school-aged children: A systematic review. *Neuroscience and Biobehavioral Reviews, 136*, 104623.
13. Chang, Y. C., Chen, W. T., Su, S. H., et al. (2022). PM(2.5) exposure and incident attention-deficit/hyperactivity disorder during the prenatal and postnatal periods: A birth cohort study. *Environmental Research, 214*, 113769.
14. Collaborators., G.N.S.D. (2024). Global, regional, and national burden of disorders affecting the nervous system, 1990-2021: A systematic analysis for the global burden of disease study 2021. *Lancet Neurology, 23*, 344–381.
15. Cortese, S., & Coghill, D. (2018). Twenty years of research on attention-deficit/hyperactivity disorder (ADHD): Looking back, looking forward. *Evidence-Based Mental Health, 21*, 173–176.
16. Cory-Slechta, D. A., Merrill, A., & Sobolewski, M. (2023). Air pollution-related neurotoxicity across the life span. *Annual Review of Pharmacology and Toxicology, 63*, 143–163.
17. Costa, L. G., Cole, T. B., Dao, K., et al. (2020). Effects of air pollution on the nervous system and its possible role in neurodevelopmental and neurodegenerative disorders. *Pharmacology & Therapeutics, 210*, 107523.
18. Cotter, D. L., Campbell, C. E., Sukumaran, K., et al. (2023). Effects of ambient fine particulates, nitrogen dioxide, and ozone on maturation of functional brain networks across early adolescence. *Environment International, 177*, 108001.
19. Cserbik, D., Chen, J. C., McConnell, R., et al. (2020). Fine particulate matter exposure during childhood relates to hemispheric-specific differences in brain structure. *Environment International, 143*, 105933.
20. Dutheil, F., Comptour, A., Morlon, R., et al. (2021). Autism spectrum disorder and air pollution: A systematic review and meta-analysis. *Environmental Pollution, 278*, 116856.
21. Essers, E., Binter, A. C., Neumann, A., et al. (2023). Air pollution exposure during pregnancy and childhood, APOE ε4 status and Alzheimer polygenic risk score, and brain structural morphology in preadolescents. *Environmental Research, 216*, 114595.
22. Faraone, S. V., Perlis, R. H., Doyle, A. E., et al. (2005). Molecular genetics of attention-deficit/hyperactivity disorder. *Biological Psychiatry, 57*, 1313–1323.
23. Guxens, M., Lubczyńska, M. J., Muetzel, R. L., et al. (2018). Air pollution exposure during fetal life, brain morphology, and cognitive function in school-age children. *Biological Psychiatry, 84*, 295–303.
24. Health Effects Institute. (2024). *State of global air 2024*. Health Effects Institute.
25. Herting, M. M., Bottenhorn, K. L., & Cotter, D. L. (2024). Outdoor air pollution and brain development in childhood and adolescence. *Trends in Neurosciences, 47*, 593–607.
26. Hertz-Picciotto, I., Schmidt, R. J., & Krakowiak, P. (2018). Understanding environmental contributions to autism: Causal concepts and the state of science. *Autism Research, 11*, 554–586.
27. Jiang, C. C., Lin, L. S., Long, S., et al. (2022). Signalling pathways in autism spectrum disorder: Mechanisms and therapeutic implications. *Signal Transduction and Targeted Therapy, 7*, 229.
28. Klocke, C., Allen, J. L., Sobolewski, M., et al. (2018). Exposure to fine and ultrafine particulate matter during gestation alters postnatal oligodendrocyte maturation, proliferation capacity, and myelination. *Neurotoxicology, 65*, 196–206.
29. Klocke, C., Allen, J. L., Sobolewski, M., et al. (2017). Neuropathological consequences of gestational exposure to concentrated ambient fine and ultrafine particles in the mouse. *Toxicological Sciences, 156*, 492–508.
30. Li, K., Li, L., Cui, B., et al. (2018). Early postnatal exposure to airborne fine particulate matter induces autism-like phenotypes in male rats. *Toxicological Sciences, 162*, 189–199.
31. Lin, L. Z., Zhan, X. L., Jin, C. Y., et al. (2022). The epidemiological evidence linking exposure to ambient particulate matter with neurodevelopmental disorders: A systematic review and meta-analysis. *Environmental Research, 209*, 112876.

32. Liu, H., Ding, L., Qu, G., et al. (2023). Particulate matter exposure during pregnancy and infancy and risks of autism spectrum disorder in children: A systematic review and meta-analysis. *Science of the Total Environment, 855*, 158830.
33. Lubczyńska, M. J., Muetzel, R. L., El Marroun, H., et al. (2020). Exposure to air pollution during pregnancy and childhood, and white matter microstructure in preadolescents. *Environmental Health Perspectives, 128*, 27005.
34. Lubczyńska, M. J., Muetzel, R. L., El Marroun, H., et al. (2021). Air pollution exposure during pregnancy and childhood and brain morphology in preadolescents. *Environmental Research, 198*, 110446.
35. Maher, B. A., Ahmed, I. A., Karloukovski, V., et al. (2016). Magnetite pollution nanoparticles in the human brain. *Proceedings of the National Academy of Sciences of the United States of America, 113*, 10797–10801.
36. Markevych, I., Tesch, F., Datzmann, T., et al. (2018). Outdoor air pollution, greenspace, and incidence of ADHD: A semi-individual study. *Science of the Total Environment, 642*, 1362–1368.
37. Min, J. Y., & Min, K. B. (2017). Exposure to ambient PM(10) and NO(2) and the incidence of attention-deficit hyperactivity disorder in childhood. *Environment International, 99*, 221–227.
38. Mortamais, M., Pujol, J., Martínez-Vilavella, G., et al. (2019). Effects of prenatal exposure to particulate matter air pollution on corpus callosum and behavioral problems in children. *Environmental Research, 178*, 108734.
39. Mortamais, M., Pujol, J., van Drooge, B. L., et al. (2017). Effect of exposure to polycyclic aromatic hydrocarbons on basal ganglia and attention-deficit hyperactivity disorder symptoms in primary school children. *Environment International, 105*, 12–19.
40. Nephew, B. C., Nemeth, A., Hudda, N., et al. (2020). Traffic-related particulate matter affects behavior, inflammation, and neural integrity in a developmental rodent model. *Environmental Research, 183*, 109242.
41. Patten, K. T., González, E. A., Valenzuela, A., et al. (2020). Effects of early life exposure to traffic-related air pollution on brain development in juvenile Sprague-Dawley rats. *Translational Psychiatry, 10*, 166.
42. Pérez-Crespo, L., Kusters, M. S. W., López-Vicente, M., et al. (2022). Exposure to traffic-related air pollution and noise during pregnancy and childhood, and functional brain connectivity in preadolescents. *Environment International, 164*, 107275.
43. Peterson, B. S., Bansal, R., Sawardekar, S., et al. (2022). Prenatal exposure to air pollution is associated with altered brain structure, function, and metabolism in childhood. *Journal of Child Psychology and Psychiatry, 63*, 1316–1331.
44. Peterson, B. S., Rauh, V. A., Bansal, R., et al. (2015). Effects of prenatal exposure to air pollutants (polycyclic aromatic hydrocarbons) on the development of brain white matter, cognition, and behavior in later childhood. *JAMA Psychiatry, 72*, 531–540.
45. Posner, J., Polanczyk, G. V., & Sonuga-Barke, E. (2020). Attention-deficit hyperactivity disorder. *Lancet, 395*, 450–462.
46. Pujol, J., Fenoll, R., Macià, D., et al. (2016a). Airborne copper exposure in school environments associated with poorer motor performance and altered basal ganglia. *Brain and Behavior: A Cognitive Neuroscience Perspective, 6*, e00467.
47. Pujol, J., Martínez-Vilavella, G., Macià, D., et al. (2016b). Traffic pollution exposure is associated with altered brain connectivity in school children. *NeuroImage, 129*, 175–184.
48. Qi, Y., Wei, S., Xin, T., et al. (2022). Passage of exogeneous fine particles from the lung into the brain in humans and animals. *Proceedings of the National Academy of Sciences of the United States of America, 119*, e2117083119.
49. Shang, L., Yang, L., Yang, W., et al. (2020). Effects of prenatal exposure to NO(2) on children's neurodevelopment: A systematic review and meta-analysis. *Environmental Science and Pollution Research International, 27*, 24786–24798.
50. Suades-González, E., Gascon, M., Guxens, M., et al. (2015). Air pollution and neuropsychological development: A review of the latest evidence. *Endocrinology, 156*, 3473–3482.

51. Sukumaran, K., Cardenas-Iniguez, C., Burnor, E., et al. (2023). Ambient fine particulate exposure and subcortical gray matter microarchitecture in 9- and 10-year-old children across the United States. *iScience, 26*, 106087.
52. Thygesen, M., Holst, G. J., Hansen, B., et al. (2020). Exposure to air pollution in early childhood and the association with attention-deficit hyperactivity disorder. *Environmental Research, 183*, 108930.
53. Woodward, N. C., Haghani, A., Johnson, R. G., et al. (2018). Prenatal and early life exposure to air pollution induced hippocampal vascular leakage and impaired neurogenesis in association with behavioral deficits. *Translational Psychiatry, 8*, 261.
54. World Health Organization. (2018). *More than 90% of the world's children breathe toxic air every day*. World Health Organization.
55. Xu, G., Strathearn, L., Liu, B., et al. (2019). Prevalence and treatment patterns of autism spectrum disorder in the United States, 2016. *JAMA Pediatrics, 173*, 153–159.
56. Yu, W., Ye, T., Zhang, Y., et al. (2023). Global estimates of daily ambient fine particulate matter concentrations and unequal spatiotemporal distribution of population exposure: A machine learning modelling study. *Lancet Planetary Health, 7*, e209–e218.
57. Yuchi, W., Brauer, M., Czekajlo, A., et al. (2022). Neighborhood environmental exposures and incidence of attention deficit/hyperactivity disorder: A population-based cohort study. *Environment International, 161*, 107120.
58. Zeidan, J., Fombonne, E., Scorah, J., et al. (2022). Global prevalence of autism: A systematic review update. *Autism Research, 15*, 778–790.
59. Zhao, J., He, T., Wang, F., et al. (2024). Association of prenatal and postnatal exposure to air pollution with clinically diagnosed attention deficit hyperactivity disorder: A systematic review. *Frontiers in Public Health, 12*, 1396251.

Chapter 6
Air Pollution as a Risk Factor of Neurodegenerative Diseases and Stroke: Mechanisms of Action and Implications

Juliana C. S. Chaves, Ian W. Peall, Emily R. Vivian, and Anthony Robert White

Keywords Air pollution · Brain health · Neurodegenerative diseases · Stroke · Particulate matter

Abbreviations

AD	Alzheimer's Disease
ADHD	Attention-Deficit Hyperactivity Disorder
ANS	Autonomic Nervous System
ASD	Autism Spectrum Disorder
AQP4	Aquaporin-4
BBB	Blood-Brain Barrier
CNS	Central Nervous System
CO	Carbon Monoxide
CVD	Cardiovascular Disease
HRs	Hazard Ratios
IDEAS	Imaging Dementia, Evidence for Amyloid Scanning
IS	Ischemic Stroke
Nrf2	NF-E2 p45-Related Factor 2
NO_2	Nitrogen Dioxide
NO_x	Nitrogen Oxides

J. C. S. Chaves (✉) · E. R. Vivian · A. R. White
QIMR Berghofer Medical Research Institute, Cellular and Molecular Neurodegeneration, Brain and Mental Health Program, Brisbane, QLD, Australia

School of Biomedical Sciences, The University of Queensland, Brisbane, QLD, Australia
e-mail: Juliana.DaSilvaChaves@qimrberghofer.edu.au

I. W. Peall
QIMR Berghofer Medical Research Institute, Cellular and Molecular Neurodegeneration, Brain and Mental Health Program, Brisbane, QLD, Australia

K. M. Kanninen, A. R. White (eds.), *Air Pollution and the Brain*,
https://doi.org/10.1007/978-3-031-99302-2_6

O_3	Ozone
PET	Positron Emission Tomography
PM	Particulate Matter
PM10	Particulate Matter (≤10 micrometers)
PM2.5	Fine Particulate Matter (≤2.5 micrometers)
POPs	Persistent Organic Pollutants
PD	Parkinson's Disease
ROS	Reactive Oxygen Species
SO_2	Sulfur Dioxide
UFPs	Ultrafine Particles
WHO	World Health Organization

6.1 Introduction

Air pollution is increasingly recognised as a serious threat not only to respiratory and cardiovascular health but also to brain function, with emerging research linking exposure to air pollutants with a range of neurological conditions [1–4]. Recently, studies have shown evidences that highlights the connection between fine particulate matter (PM2.5), nitrogen dioxide (NO_2), carbon monoxide (CO), sulfuric dioxide (SO_2), and ozone (O_3) with cognitive decline, stroke, and neurodegenerative diseases such as Alzheimer's (AD) and Parkinson's disease (PD) [1–3, 5].

Interestingly, it was reported that long-term exposure to elevated levels of PM2.5 exceeding the World Health Organization (WHO) standards is associated with significantly increased risks of mortality and hospital admissions for AD, non-AD dementia, and PD in elderly populations [6]. Specifically, residents in areas with high PM2.5 levels had higher age-adjusted rates of mortality and hospital admissions for AD, non-AD dementia, and PD compared to those in areas with low levels of air particulate matter [6]. The risk of death from AD was significantly higher in areas with elevated PM2.5 levels, while the effects on non-AD dementia and PD were less pronounced [6]. Further research has explored the relationship between neurodegenerative disorders, such as AD and PD, and the glymphatic system, a mechanism integral to the clearance of metabolic waste from the brain [7–9]. Dysregulation in the glymphatic system, partly regulated by aquaporin-4 (AQP4), has been associated with protein accumulation in AD and PD [8]. Lifestyle factors such as sleep deprivation and exercise can influence glymphatic system function, potentially affecting the pathogenesis of AD and PD [8].

Additionally, neuroinflammation has been identified as a key player in the development of AD and PD, commonly associated with protein accumulation and activation of microglia, leading to synaptic damage and neuronal loss [7, 8, 10]. The transcription factor NF-E2 p45-related factor 2 (Nrf2) has emerged as a potential therapeutic target for its role in regulating redox homeostasis and anti-inflammatory functions in neurodegenerative diseases [7]. Activation of Nrf2 has shown promise in reducing inflammation, improving mitochondrial function, and potentially serving as a novel therapeutic approach for addressing the pathogenesis of AD and PD [7]. In addition, other studies have shown that both short-term and long-term exposure to these pollutants can accelerate cognitive impairment and contribute to brain infarcts, with severe implications for public health [11–13].

Recent research has also drawn attention to the alarming association between air pollution and stroke, particularly ischemic strokes. Various studies have demonstrated that pollutants like PM2.5, PM10, NO_2, and SO_2 are strongly linked to increased incidence and mortality rates from stroke [5, 12]. This association is especially pronounced in younger individuals, those with pre-existing health conditions like hyperlipidaemia, and in specific seasonal periods, such as autumn and winter [14, 15]. Additionally, long-term exposure to air pollution is shown to elevate the risk of ischemic stroke (IS) across diverse populations, suggesting that even subtle environmental pollutants can have significant long-term effects on neurological health [11, 16] (Fig. 6.1).

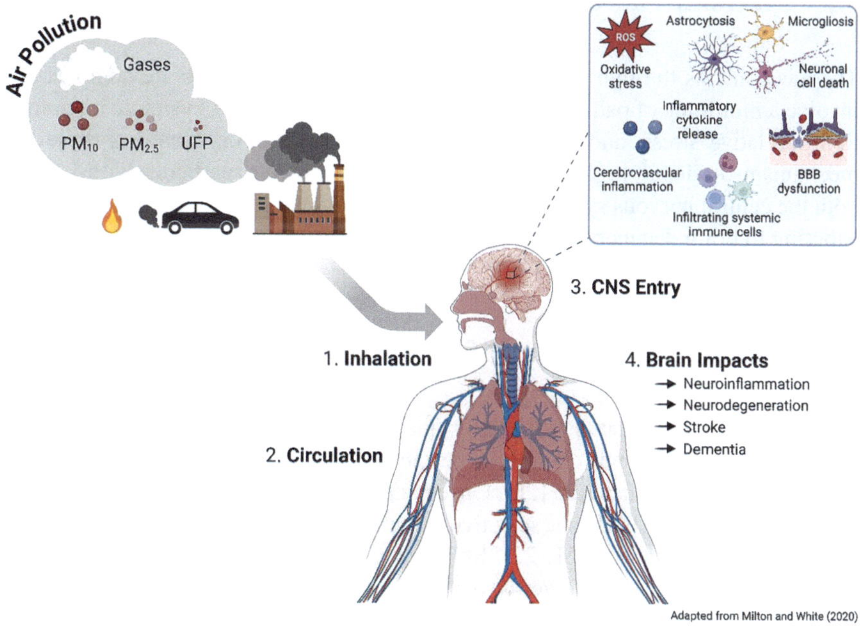

Fig. 6.1 Pathways of air pollution entry and impact on the human brain. This figure illustrates how air pollution affects the central nervous system (CNS) through four step pathway: (1) inhalation of pollutants like particulate matter (PM), nitrogen dioxide (NO_2), carbon monoxide (CO), and volatile organic compounds (VOCs) via the respiratory system; (2) transport through the circulatory system, where pollutants enter the bloodstream and may reach the brain; (3) crossing the blood-brain barrier (BBB), especially fine particulate matter (PM2.5), which can penetrate or bypass the barrier; (4) impact on the CNS, causing inflammation, oxidative stress, and potential cognitive and mental health disorders. (*Source:* Figure adapted from Milton and White, 2020, and created using BioRender.com [17])

The harmful effects of air pollution on brain health and its association with an increased risk of stroke have led to widespread calls for immediate public health interventions. Reducing concentrations of airborne pollutants, such as PM10, NO_2, and SO_2, has the potential to substantially reduce the incidence of stroke and cognitive decline, providing a significant opportunity to enhance brain health globally. Understanding the complex relationship between air pollution and brain function is essential for the development of targeted policies and strategies aimed at mitigating these risks, particularly for vulnerable populations. This chapter explores the expanding body of evidence linking air pollution to adverse brain health outcomes, with a specific emphasis on cognitive decline and stroke. By examining recent research, we seek to uncover the underlying mechanisms driving these associations and underscore the urgent need for proactive measures to reduce environmental pollutants, ultimately preventing neurological diseases and improving public health on a global scale.

6.2 Mechanisms of Action of Air Pollution on Brain Health

The mechanisms through which air pollution affects brain health are complex and involve a multitude of pathways, including direct neurotoxicity, systemic inflammation, oxidative stress, and vascular dysfunction. This section examines these key mechanisms in detail, highlighting how airborne pollutants, particularly PM, affect both the central nervous system (CNS) and vascular health [18, 19], ultimately contributing to brain degeneration.

6.2.1 Direct Neurotoxicity of Air Pollutants

Airborne particulate matter, especially fine particles such as PM2.5 and ultrafine particles UFPs, are small enough to be inhaled deeply into the lungs and subsequently enter the bloodstream [20]. Once in circulation, these particles can cross the blood-brain barrier (BBB), a selective membrane that usually protects the brain from harmful substances [21, 22]. Studies have demonstrated that these particles can infiltrate brain tissue, triggering a cascade of inflammatory responses and direct cellular damage [1, 2, 9, 23]. Once in the brain, particulate matter interacts with microglial cells, the resident immune cells of the CNS [23]. The activation of microglia leads to the release of pro-inflammatory cytokines and reactive oxygen species (ROS), both of which contribute to neuroinflammation and neuronal damage [22, 23]. Chronic activation of microglia is a key process in the pathogenesis of neurodegenerative diseases such as AD and PD.

6.2.2 Inflammatory Pathways and Oxidative Stress

In addition to direct neurotoxicity, air pollution promotes systemic inflammation and oxidative stress, both of which have significant effects on brain health. Exposure to air pollutants such as PM2.5 and UFPs triggers inflammatory responses in various tissues, including the brain. Inflammation is a central mechanism linking air pollution to neurological damage [24, 25]. Particulate matter, particularly UFPs, can cross biological barriers, such as the BBB, and activate immune cells in the brain, leading to the production of pro-inflammatory cytokines [26, 27] (Fig.6.2). These cytokines, such as interferon-gamma, are elevated not only in brain tissue but also in peripheral tissues, as seen in animal models [28, 29]. Chronic inflammation in the brain can exacerbate the progression of neurodegenerative conditions, including AD. The sustained inflammatory environment damages neurons and synapses, impairing cognitive function over time [25].

In parallel with inflammation, air pollution also contributes to oxidative stress, which is another key pathway through which pollutants affect brain health [18]. Exposure to pollutants such as PM2.5 and UFPs results in the generation of ROS, which are highly reactive molecules that cause damage to cellular components, including lipids, proteins, and DNA [4, 30]. This oxidative stress triggers mitochondrial dysfunction, elevating oxidative stress markers and exacerbating inflammation in the brain [31]. The combined effects of inflammation and oxidative stress lead to neuronal injury, synaptic dysfunction, and ultimately neurodegeneration, particularly in brain regions involved in memory and cognition, such as the hippocampus.

6.2.3 Endothelial Dysfunction and Vascular Pathways

Air pollution also exerts significant effects on the vascular system, contributing to both cerebrovascular dysfunction and increased stroke risk [18]. One of the primary ways in which air pollution affects vascular health is through endothelial dysfunction, as evidenced by reductions in flow-mediated dilation (FMD) and alterations in circulating levels of markers associated with endothelial health [32]. The endothelium, the thin layer of cells lining blood vessels, plays a crucial role in regulating blood flow, vascular tone, and coagulation [33, 34]. Exposure to air pollutants, including particulate matter, can disrupt the integrity of the endothelial cells, leading to increased permeability and transmigration of monocytes across the endothelial monolayer [35, 36]. This disruption in endothelial cell integrity can impair their ability to regulate blood flow, and potentially lead to a series of vascular events.

Studies have also shown that pollutants such as PM2.5 and UFPs activate the complement system, which in turn promotes inflammation and exacerbates endothelial dysfunction [4, 28]. Additionally, particulate matter exposure has been shown to upregulate pro-inflammatory cytokines in endothelial cells, further disrupting the BBB and leading to systemic inflammation [37, 38]. These vascular changes

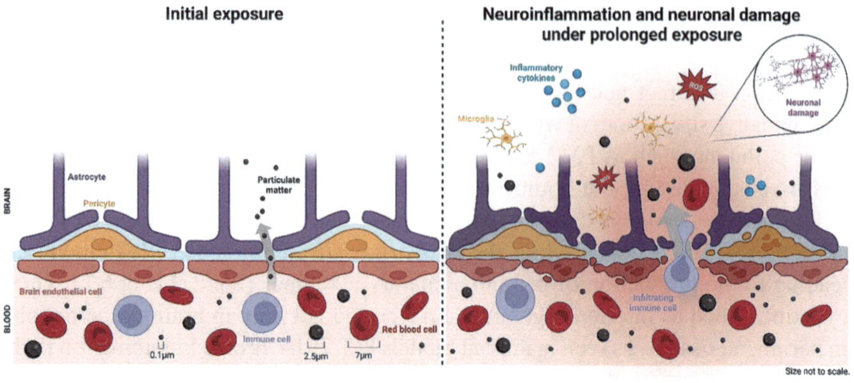

Fig. 6.2 Proposed effect of prolonged exposure to air pollutants and blood-brain barrier (BBB) dysfunction. This figure emphasises the critical role of the BBB in regulating the brain's response to prolonged exposure to airborne pollutants and its subsequent neurotoxic effects. Air pollutants trigger an inflammatory response in the brain, leading to a cascade of events that include activation of microglia and astrocytes, the release of pro-inflammatory cytokines, and the production of reactive oxygen species (ROS), all of which contribute to disruption of tight junctions and endothelial cells, essential for maintaining BBB integrity [40]. (Figure was created using BioRender.com)

contribute to atherosclerosis, or the build-up of fatty plaques in the arteries, which in turn increases the risk of stroke, particularly ischemic stroke. The combination of endothelial injury, impaired blood flow regulation, and heightened inflammation results in a higher incidence of cerebrovascular events in individuals exposed to air pollution.

Moreover, the effects of air pollution on endothelial integrity are not limited to the arteries supplying the brain. Pollutants like PM2.5 and SO_2 can induce systemic inflammation, which further impacts vascular health throughout the body [39]. This chronic inflammatory state increases the risk of cardiovascular diseases, including heart attack and stroke, by promoting the formation of blood clots and plaques that obstruct blood vessels (Fig. 6.2).

6.2.4 Disruption of Neurodevelopmental Processes

In addition to its effects on adults, air pollution poses significant risks to the developing brain. Studies have found that children exposed to high levels of air pollution during critical periods of neurodevelopment are at an increased risk for cognitive and behavioural impairments [41, 42]. The developing brain is particularly vulnerable to the toxic effects of air pollutants, which can interfere with processes like neurogenesis, synaptic plasticity, and myelination [41]. Disruption of these processes can lead to delayed cognitive development and an increased risk of neurodevelopmental disorders such as attention-deficit hyperactivity disorder (ADHD) and autism spectrum disorder (ASD). Long-term exposure to air pollution in childhood may also contribute to learning disabilities and other cognitive deficits later in life.

6.2.5 Hormonal and Metabolic Effects

Emerging evidence also suggests that air pollution may influence brain health through hormonal and metabolic pathways [2]. Recent studies have shown that pollutants such as NO_2 and PM2.5 alter levels of stress hormones like cortisol, which are critical for regulating mood, memory, and overall cognitive function [1, 43]. Chronic exposure to air pollution can elevate cortisol levels, leading to long-term stress and potential damage to brain regions involved in memory and emotion regulation [1, 43]. Elevated cortisol has been linked to neurodegenerative processes, including hippocampal atrophy, a common finding in the pathology of AD. In addition to hormonal changes, air pollution has been implicated in the development of metabolic disorders, such as insulin resistance, which is a known risk factor for AD [44]. Exposure to pollutants triggers systemic inflammation and oxidative stress, both of which contribute to impaired glucose metabolism [44, 45]. Insulin resistance, in turn, affects brain function, as glucose is the primary energy source for neurons. Disrupted glucose metabolism has been shown to impair synaptic function, reduce neuroplasticity, and contribute to the cognitive decline associated with neurodegenerative diseases, like AD [44].

6.3 Epidemiological Evidence of Air Pollution on Brain Health

6.3.1 Studies and Statistics in Neurodegenerative Diseases

Numerous epidemiological studies have suggested a potential link between air pollution, specifically PM and NO_2 from traffic-related sources, and neurodegenerative diseases, including dementia, AD, PD, and Parkinsonism [46, 47]. Existing evidence indicates that chronic exposure to these pollutants may result in structural changes in the brain, accumulation of tau proteins, autophagy dysfunction, and neuroinflammation, all of which are associated with the onset and progression of neurodegenerative conditions [46]. There is a critical need for further long-term studies in regions with higher pollution levels to validate these risks and identify potential preventive measures. Addressing the effects of air pollution on neurodegenerative diseases is crucial for mitigating the burden of these conditions on both individuals and society [39, 46, 47].

A cross-sectional study conducted in the United States (U.S.) investigated the relationship between ambient air pollution and the likelihood of amyloid positron emission tomography (PET) scan positivity among over 18,000 older adults with cognitive impairment [48]. The study used data from the Imaging Dementia, Evidence for Amyloid Scanning (IDEAS) Study, which included participants diagnosed with mild cognitive impairment (60.5%) or dementia (39.5%) who underwent amyloid PET scans [48]. The researchers assessed air quality data from the

Environmental Protection Agency's Downscaler model, which provided estimates of fine PM2.5 and O_3 concentrations at participants' residential locations. Two distinct time windows were analysed, 2002–2003 (approximately 14 years prior to the PET scan) and 2015–2016 (approximately 1 year prior to the PET scan), to examine both short-term and long-term exposure to air pollution. The study found that higher concentrations of PM2.5 were associated with an increased likelihood of a positive amyloid PET scan [48]. Specifically, for the 2002–2003 time period, each 4 $\mu g/m^3$ increase in PM2.5 corresponded to a 10% increase in the odds of amyloid PET positivity, after adjusting for demographic, lifestyle, socioeconomic, and medical factors. Similar associations were observed for the 2015–2016-time window [48]. In contrast, exposure to elevated O_3 concentrations was not linked to amyloid PET positivity. Marginal effects analyses revealed a dose–dependent relationship, showing that individuals living in areas with the highest PM2.5 levels had up to a 5% higher probability of having a positive amyloid PET scan compared to those in the lowest pollution areas. This association remained significant even after accounting for geographic differences. The researchers found no evidence suggesting that the relationship between PM2.5 exposure and amyloid PET positivity varied by sex or clinical stage (mild cognitive impairment vs. dementia) [48]. These findings indicate that exposure to air pollution, particularly fine PM, may be linked to the deposition of amyloid-beta, a key hallmark of AD pathology, in older adults with cognitive impairment [48]. The authors conclude that these results highlight the importance of considering airborne toxic pollutants in public health policies and individual risk assessments for AD and dementia.

Another study, using a similar approach to determine the association between the long-term air pollution exposure to ambient fine PM2.5, NO_2, and O_3 and prevalence of dementia and AD in the US adults aged 65 and older [49]. The study used data from two national population-based cohorts spanning from 2000 to 2018. Hazard ratios (HRs) for incident dementia and AD were calculated per interquartile range (IQR) increase in exposure to PM2.5, NO_2, and O_3 [49]. It was found that exposure to PM2.5 and NO_2 was associated with an increased incidence of both dementia and AD, while exposure to O_3 appeared to have a protective effect [49]. The study identified approximately 2.0 million new cases of dementia and 0.8 million new cases of AD. HRs demonstrated a positive association between PM2.5 and NO_2 with both dementia and AD, with HRs of 1.060 for dementia and 1.078 for AD per IQR increase in PM2.5, and HRs of 1.019 for dementia and 1.031 for AD per IQR increase in NO2. On the other hand, O_3 exposure showed HRs of 0.990 for dementia and 0.982 for AD, indicating a slight protective effect [49]. The concentration–response relationships for PM2.5 and NO_2 were found to be approximately linear, suggesting a direct and consistent association with increased disease incidence. The study also examined potential effect modifications by factors such as sex, race, Medicaid eligibility, age groups, and population density. It found that certain subgroups, including Black individuals, those eligible for Medicaid, and individuals under age 75, had higher hazard ratios for dementia and AD. Sensitivity analyses confirmed the robustness of the associations between PM2.5 and NO_2 exposure and the incidence of dementia and AD. In addition, the study's estimates of attributable fraction suggested that reductions in PM2.5 and NO_2 exposure could potentially decrease the incidence of dementia and AD in the USA. This provides compelling evidence that long-term exposure to PM2.5 and NO_2 is associated with an increased risk of dementia and AD in older adults [49]. While O_3 exposure demonstrated a protective effect, the linear concentration–response relationships observed for PM2.5 and NO_2 suggest a consistent, harmful impact on cognitive health. Altogether, these findings underscore the importance of air quality interventions, particularly the reduction of PM2.5 and NO_2 levels, as potential strategies to mitigate the incidence of these neurodegenerative diseases in the elderly population.

A study conducted in 2023 by Krzyzanowski identified 346 PD patients in Olmsted County, Minnesota, between 1991 and 2015, who were matched with 4813 controls. Exposure to PM2.5 and NO_2 was assessed based on the residential address history of participants, covering up to 10 years prior to the onset of PD symptoms. The study's findings showed that higher levels of PM2.5 exposure were linked to an increased risk of developing PD [50]. Compared to the lowest quintile of PM2.5 exposure, the risk of PD was 4% higher in the second quintile and increased by 14% in the highest quintile. Furthermore, elevated NO_2 exposure was associated with a greater risk of PD, with a 5% increased risk in the fourth quintile and a 13% increase in the top quintile. In addition, the study demonstrated that, among PD patients, higher PM2.5 exposure was associated with a 36% increased risk of developing the akinetic-rigid subtype of PD compared to the tremor-predominant subtype [50]. Both PM2.5 and NO_2 exposure were linked to a greater risk of developing dyskinesia in PD patients, with a 42% increase in risk per 1 μg/m^3 increase in PM2.5 [50]. However, PM2.5 exposure was not associated with all-cause mortality in PD patients. These findings suggest that reducing air pollution, particularly PM2.5, may help mitigate the risk of developing PD, alter the clinical presentation of the disease, and reduce the likelihood of dyskinesia in PD patients [50]. This highlights the potential role of environmental exposures, such as air pollution, in the pathogenesis and clinical progression of PD.

6.3.2 *Studies and Statistics in Stroke*

Chronic exposure to air pollution has been linked to an increased risk of ischemic stroke, a type of neurovascular disease [51, 52]. The diversity of air pollution's effects on ischemic stroke risk is influenced by factors such as the type of pollutant, duration and intensity of exposure, individual susceptibility, and geographic variations [53]. Long-term exposure may not only increase the incidence of stroke, but also worsen post-stroke outcomes, contributing to cognitive decline and neurological impairments [43]. Understanding these complex interactions is essential for developing targeted public health interventions to mitigate the long-term impact of air pollution on stroke morbidity and mortality.

A study conducted in 2021 by Wolf et al. investigated long-term exposure to low-level ambient air pollution across Europe and the incidence of stroke and coronary heart disease. Specifically, they looked at the effects of fine PM2.5, NO_2, black carbon, and warm-season ozone (O_3) on the increased risk of stroke and coronary heart disease across six pooled populations from Sweden, Denmark, the Netherlands, and Germany. During the study, data on 137,148 participants were analysed, and the findings indicated that the incidence of stroke was associated with PM2.5, NO_2, and black carbon, while coronary heart disease incidence was only associated with NO_2. Conversely, O_3 was not associated with an increase in either outcome. Concentration-response curves indicated no evidence of a threshold below which air pollutant concentrations are not harmful to cardiovascular health. The study's analyses remained elevated even when restricting analyses to participants exposed to pollutant concentrations lower than the EU limit values of 25 μg/m^3 for PM2.5 and 40 μg/m^3 for NO_2. Despite some limitations, such as potential exposure misclassification and limited consideration of additional contextual factors, the study provides crucial evidence to support the associations between long-term PM2.5, NO_2, and black carbon exposure with incident stroke and coronary heart disease at exposure concentrations lower than current regulatory limits [54].

A similar study conducted by Stockfelt et al investigated the effects of PM air pollutants on incident cardiovascular disease for different time windows of exposure. The study utilised two Swedish cohorts, the PPS cohort and the GOT-MONICA cohort, and followed participants from 1990 to 2011. The data on residential addresses and assigned yearly total and source-specific PM and NO_2 exposures were collected based on dispersion models. The study found that long-term exposure to PM air pollutants was associated with ischemic heart disease and stroke, particularly in women, non-smokers, and high socio-economic classes at relatively low exposure levels. Interestingly, these associations were much higher in the last 5 years of the study; however, they could not be attributed to a specific source of PM air pollutant [55].

Ma et al. conducted a large, nationwide population-based cohort study that investigated the impact of long-term exposure to ambient air pollutants on an elderly population in the United States. The study used Medicare data for over 17 million fee-for-service beneficiaries aged 65 and older between 2000 and 2017, combined with high-resolution air pollution data across the contiguous USA. The results showed that increased levels of air pollutants were associated with a higher risk of stroke incidence. An interquartile range (IQR) increase in the annual average NO_2 (12 ppb) preceding 1-year diagnosis was associated with a 7.3% increase in risk of stroke in the single air pollutant model, with the effect size changing little in models adjusting for other pollutants. An IQR increase in annual average PM2.5 (3.7 μg/m^3) was associated with a 5.6% increased risk in the single pollutant model, dropping 2.2% in the tri-pollutant model. An IQR increase in average warm-season O_3 (6.5 ppb) was associated with a 3.5% increased risk in the single-pollutant model, dropping to 2.1% in the tri-pollutant model. The study established the strongest relationship between NO_2 and stroke risk, suggesting that traffic-related air pollution may play a particularly significant role. The study also suggests that if NO_2 levels could be lowered by the IQR of 12.4 ppb, 6% of stroke cases could be avoided [56].

A study by Fisher et al. investigated the association between short-term exposure to PM air pollution and the risk of ischemic and haemorrhagic stroke. They utilised a case-crossover design and a nationwide kriging model to examine short-term ambient exposure to PM10 and PM2.5 on participants of the Health Professionals Follow-up Study. They collected data on strokes that occurred between 1990 and 2010 and used conditional logistic regression models to obtain estimates of odds and ratios and 95% confidence intervals associated with an IQR increase in PM2.5 and PM10. They found a positive statistically significant association between PM10 and ischemic stroke, particularly among non-smokers, aspirin non-users, and those with a history of high cholesterol. However, the study found no evidence of a positive association between short-term exposure to PM and haemorrhagic stroke or between PM2.5 and ischemic stroke [57].

Another recent study in 2023 by Keller et al. assessed the impact of long-term air pollution exposure on ischemic stroke mortality in Germany from 2015 to 2019. The study used data from the German Nationwide Inpatient Sample and the German Federal Environmental Agency to analyse the association between air pollution parameters and in-hospital case-fatality of ischemic stroke patients. The study included over 1.5 million hospitalisations of patients with ischemic stroke in Germany. The key findings of the study showed that elevated long-term air pollution levels in residential areas of benzene, O_3, NO, SO_2, and PM2.5 were significantly associated with increased case-fatality independent of age, sex, cardiovascular risk factors, comorbidities, and revascularisation treatments. Additionally, SO_2 concentrations were significantly associated with a stroke-case fatality rate of over 8%, independent of residence area type and area use. The study also compared patients residing in federal districts with high versus low long-term air pollution exposure and found associations between high air pollution concentrations and increased ischemic stroke mortality. Specifically, long-term exposure to benzene, O_3, NO, and PM2.5 was independently associated with an increased in in-hospital case-fatality of ischemic stroke patients [58].

These studies highlight the urgent need to reduce exposure to air pollution through emission control strategies to reduce stroke burden and stroke mortality.

6.4 Population Vulnerabilities

Although the effects of long-term exposure to air pollution are not limited to one particular demographic, there are some populations that are more susceptible:

1. *Aged individuals:* Studies have indicated that elderly individuals are particularly vulnerable to the health effects of air pollution [59]. Studies conducted in China found that increasing levels of air pollution resulted in negative health effects on the elderly, leading to significantly higher rates of hospital visits among this population [60, 61]. Additionally, research in other countries suggests that chronic exposure to air pollution may contribute to respiratory diseases such as

chronic obstructive pulmonary disease (COPD) among older adults, they are at higher risk of developing chronic diseases like cardiovascular disease, hypertension, and respiratory conditions, all of which can be exacerbated by long-term exposure to air pollution [62, 63].

2. *Individuals with pre-existing conditions:* People with respiratory diseases (eg., asthma, chronic obstructive pulmonary disease) or cardiovascular diseases (eg., hypertension, atherosclerosis) are potentially more susceptible to the inflammatory and oxidative stress caused by long-term air pollution exposure. Air pollution can worsen their condition and increase the likelihood of severe outcomes such as strokes or heart attacks [64–68].
3. *Low socioeconomic status (SES) populations:* Individuals from lower SES backgrounds are often disproportionately exposed to higher levels of air pollution due to living near industrial areas, high-traffic zones, or densely populated urban centres. Additionally, they may also have limited access to healthcare resources, making it harder to manage health conditions that air pollution exacerbates. Poor nutrition and housing conditions can further compound their vulnerability [69].

As the global population ages, the risk of chronic diseases, including ischemic stroke, AD and PD, is rising sharply, placing immense pressure on healthcare systems worldwide. According to the World Health Organisation (WHO), by 2050, the proportion of people aged 60 and above is expected to double, with many countries, particularly in low- and middle-income regions, facing a rapidly aging demographic [70]. This shift significantly amplifies the threat posed by long-term air pollution, as elderly populations are more vulnerable to its effects. Air pollution accelerates vascular aging through mechanisms like systemic inflammation, oxidative stress, and endothelial dysfunction, compounding the age-related decline in cardiovascular health [71]. However, many healthcare systems are ill-equipped to handle the mass hospitalisations and long-term care that accompany an increase in dementia and stroke incidence in aging populations. Any neurological conditions, including dementia or stroke patients, often require extensive rehabilitation, specialised care, and ongoing management of neurological impairments, straining already overburdened healthcare infrastructure. In many countries, there is a shortage of healthcare professionals trained in geriatric care, limited their rehabilitation facilities, and insufficient resources to provide comprehensive care for the elderly [72, 73]. Moreover, the financial burden on both healthcare systems and families is growing as stroke recovery often involves long-term care, which many countries are not financially or logistically prepared to support [74].

6.5 Policy Recommendations

Alleviating long-term exposure to air pollution requires a combination of policy, technological innovation, and community-based interventions aimed at reducing pollution sources and minimising public exposure.

1. *Strengthening environmental policies and regulations:* Governments can adopt and enforce stricter air quality standards to limit emissions from key sources such as industrial facilities, power plants, and vehicles. Regulations targeting specific pollutants, like PM2.5, NO_2, and SO_2, can significantly reduce overall pollution levels. Initiatives like the Clean Air Act in the USA and similar frameworks in other countries have shown that stronger regulations can improve air quality and public health [75].
2. *Promoting clean energy and transportation:* Transitioning from fossil fuels to renewable energy sources (e.g., solar, wind, and hydropower) is essential for reducing air pollution. Encouraging the use of electric vehicles and expanding public transportation networks can also significantly reduce emissions from one of the largest contributors to air pollution, transportation. Governments can offer incentives for clean energy adoption, such as tax breaks or subsidies for EV purchases, while phasing out subsidies for fossil fuels [76–78].
3. *Urban planning and green infrastructure:* Well-planned urban design that includes more green spaces, such as parks and urban forests, can help absorb pollutants and improve air quality. Trees and plants act as natural filters by capturing particulate matter and releasing oxygen, creating cleaner air in urban areas. Promoting compact, walkable cities can also reduce reliance on cars and lower transportation-related emissions [75].
4. *Improving air filtration and ventilation:* For vulnerable populations, especially in heavily polluted areas, indoor air quality improvements can reduce exposure. Installing high-efficiency particulate air (HEPA) filters in homes, schools, and workplaces can help mitigate the effects of outdoor pollution. Ensuring proper ventilation in buildings, particularly in high-traffic or industrial zones, is also crucial for reducing the impact of pollutants indoors.
5. *Encouraging behavioural changes:* Public health campaigns that raise awareness about the dangers of long-term air pollution exposure can encourage individuals to adopt protective behaviours. This includes limiting outdoor activities during high pollution days, using masks, and investing in air purifiers. Reducing household emissions by promoting cleaner cooking fuels and energy-efficient appliances can also reduce individual contributions to pollution [79].
6. *Investing in research and monitoring:* Continued research into the health impacts of air pollution and the development of advanced monitoring networks allows governments to provide real-time pollution data and issue health advisories, helping communities to better understand and mitigate their exposure. Advanced analytics and machine learning can help identify pollution hotspots and model future trends for more proactive measures [80–82].
7. *Global cooperation and climate action:* Since air pollution is a transboundary issue, international cooperation is essential for tackling it on a global scale. Participation in global agreements like the Paris Climate Accord can lead to coordinated efforts to reduce emissions, particularly of pollutants that contribute to both climate change and air quality deterioration. Sharing knowledge and technologies between nations can help less developed countries implement effective pollution reduction strategies [83].

6.6 Public Awareness

At present, there is a disconnect between Scientific Research Institutes and the public in communicating relevant data and results. Promoting public awareness on these issues is becoming more and more complex, particularly with the rise of social media. There is often confusion or misinformation about the sources, effects, and solutions related to air pollution. Misinformation campaigns can also downplay its seriousness, complicating efforts to promote accurate information [84]. This misinformation is compounded by a lack of immediate health impact due to most effects being long-term, with many people not experiencing immediate symptoms. Air pollution itself is derived from multiple different sources, such as industrial, vehicular, and agricultural emissions. Which means enacting drastic change in these areas will more than likely cost people employment, which makes them less likely to participate in active change [85]. Additionally, the case that the major contributors to air pollution often have significant influence over policy and public discourse, sometimes downplaying the issue to protect their economic interests. This creates resistance to public awareness campaigns, especially if jobs or livelihoods are perceived to be at stake. Another challenge in communicating with the public is that scientific research often involves specialised terminology and complex methodologies that are difficult for non-experts to understand. Researchers are trained to communicate with precision and detail, while the public generally expects clear, simplified, and relatable explanations. Bridging this gap can lead to over simplifications that risk losing the nuance and accuracy of the findings [86]. In recent years, there has also been growing scepticism about scientific expertise, often fuelled by political or ideological views. Some members of the public distrust scientists, viewing them as part of a distant or elitist group, or even biased by funding sources or institutional affiliations. Science is also self-correcting, meaning that new data can lead to revisions or even contradictions of previous findings. This can create confusion or erode trust in scientific conclusions if the public is not accustomed to the iterative nature of scientific progress. A change in guidelines (e.g. on health or environmental policies) can be seen as a failure rather than a natural evolution of understanding [87].

6.7 Research Gaps

Existing research primarily focuses on the association between air pollution and respiratory and cardiovascular health, with limited emphasis on brain health outcomes. Current research gaps in understanding the long-term effects of air pollution on brain health and associated diseases highlight several areas where more investigation is needed to fully comprehend the risks and mechanisms involved. Longitudinal studies investigating the long-term effects of air pollution on brain health are scarce, hindering a comprehensive understanding of the impacts over extended periods. Research gaps exist in identifying specific neurobiological mechanisms underlying the effects of chronic air pollution exposure on brain health, such as neuroinflammation, oxidative stress, and BBB disruption [43, 88]. These include:

1. *Understanding the Mechanistic Pathways*
   - *Inflammation and Oxidative Stress:* While it is well-established that air pollution causes inflammation and oxidative stress, the precise molecular pathways linking these responses to neurodegenerative diseases and stroke need further elucidation. For example, the role of systemic vs localised inflammation and the precise biochemical mediators remain areas of active research [89–91].
   - *Endothelial Dysfunction:* Air pollution is thought to cause endothelial dysfunction, a key precursor to atherosclerosis; however, more research is required to understand how specific pollutants affect the endothelium over time and how this contributes to plaque formation, blood clotting, and arterial stiffness [92–95]. In addition, it is well known that endothelial dysfunction in the BBB increases its permeability, which makes the brain more vulnerable to infections and inflammation. This heightened permeability allows harmful substances, such as pathogens and inflammatory mediators, to penetrate the brain, potentially leading to neuroinflammation and further compromising neuronal health [96].
   - *Autonomic Nervous System Regulation (ANS):* There is growing evidence that air pollution can disrupt autonomic nervous system function, leading to irregular heart rates and arrhythmias, and that after stroke, it can significantly affect ANS regulation. However, the exact mechanisms through which pollutants affect the ANS, and how this contributes to cardiovascular events like sudden cardiac arrest, are not well understood [97].
2. *Role of Specific Pollutants*
   - *Particulate Matter (PM2.5) vs Ultrafine Particles:* While many studies have focused on the harmful effects of PM2.5, there is limited research on the role of ultrafine particles in cardiovascular and respiratory health. Research on the influence of ultrafine particles on brain health is limited but emerging. Ultrafine particles, which can penetrate deep into the lungs and enter the bloodstream, may have distinct and more severe cardiovascular effects. However, the long-term impacts of ultrafine particles on brain health are less well studied. Further research is needed to fully understand the specific and potentially more severe effects of ultrafine particles on brain health compared to larger PM2.5 particles [20, 98, 99].
   - *Ozone (O_3) and Nitrogen Oxides (NO_x):* Although particulate matter has been a major focus of research, other pollutants such as ozone and nitrogen oxides are less understood in the context of long-term effects on brain health. Research investigating the effects of NOx on brain health has shown a concerning link between NOx exposure and neurological disorders such as AD and PD [100]. A study using mouse models has demonstrated that NO_2

inhalation can worsen cognitive function and induce pathological abnormalities related to AD [100]. Additionally, a clinical cohort study has found associations between NOx and increased dementia risk, emphasising the importance of minimising exposure and monitoring levels in the brain [101]. Besides its relationship with dementia, research has also observed that short-term exposure to NO_2 and NO_x has a positive association with stroke, and it is more pronounced in women than in men [102]. However, further research is required to fully understand the impact of O_3 and NO*x* on brain health and cognitive function [103–105].
- *Persistent Organic Pollutants (POPs):* Emerging studies suggest that long-term exposure to POPs, including certain airborne chemicals, may contribute to cardiovascular disease, including risk of stroke, through metabolic disruption. Furthermore, other studies have demonstrated that POPs can induce oxidative stress in in vitro neuronal models, disrupt hippocampal function, and penetrate the BBB, resulting in their accumulation within brain tissue. Collectively, these disruptions contribute to the exacerbation of neurodegenerative processes, including AD and PD [106–108]. However, the long-term impacts of these pollutants remain underexplored [51, 52, 109].

3. *Exposure Timing and Duration*

   - *Cumulative Lifetime exposure:* Most of the focus of the current research assesses short- or medium-term exposure to air pollution, but less is known about the cumulative effects of lifelong exposure to low or moderate levels of pollutants. Longitudinal studies that track individuals over decades are needed to better understand the risk of developing chronic diseases after long-term, low-level exposure [110–112].
   - *Early Life Exposure:* There is increasing evidence that suggests that exposure to air pollution during early stages of embryogenesis and early childhood development can have long-term effects on brain and cardiovascular health. However, the exact mechanisms and timing of these effects, and how early life exposures might interact with adult exposure to increase neurological diseases risk requires further investigation [113–116].

4. *Subclinical Cardiovascular Events*

   - *Preclinical Markers:* Many studies have focused on the incidence of major cardiovascular events such as heart and strokes in response to air pollution exposure. However, there is a research gap in understanding how long-term air pollution exposure affects subclinical markers of cardiovascular health, such as arterial stiffness, microvascular dysfunction and heart rate variability, which often precede overt disease [117, 118].

5. *Populations at Higher Risk*

   - *Age and Sex differences:* While older adults are more susceptible to air pollution-related neurological conditions and cardiovascular diseases, there is a limited understanding to the age-specific mechanisms at play. Similarly,

sex-based differences in vulnerability to air pollution, neurodegenerative diseases and risk of stroke, such as hormonal influences or differences in occupational exposure need much more investigation [119–121].
- *Socioeconomic and Geographic Disparities:* Low-income communities and urban areas tend to experience higher levels of air pollution, and research suggests that these populations may face increased risk of develop neurodegenerative diseases and stroke [122]. However, more investigation is required to understand the interaction between socioeconomic status, access to healthcare, and long-term air pollution exposure and its impacts on brain health [88, 123–127].
- *Pre-existing Conditions:* Individuals with pre-existing conditions like diabetes, hypertension, or asthma may be more vulnerable to air pollution's cardiovascular effects including stroke, as well as increased risk of dementia, but there is a need for more focused research on how these conditions interact with long-term pollutant exposure to worsen these health outcomes [128, 129].

6. *Interaction with Other Environmental and Lifestyle Factors:*
   - *Climate Change:* As climate change exacerbates air pollution, such as through wildfires, heat waves, understanding how the combination of rising temperatures and pollution impacts brain health is a key research gap. This includes the potential synergistic effects of heat stress and air pollution on neurological conditions [130, 131].
   - *Lifestyle Factors:* There is limited research on how lifestyle factors, such as diet, exercise, and smoking, interact with air pollution to negatively effect brain health. For instance, the protective role of dietary antioxidants and the impact of exercise in polluted environments remain poorly understood [132].
7. *Intervention and Policy Impact*
   - *Effectiveness of Interventions:* Measures to control air pollution, such as reducing emissions from vehicles and industry, have been shown to improve air quality. However, a comprehensive and in-depth study is needed to evaluate the effectiveness of these policy interventions and determine which strategies provide the greatest public health benefits [133].
   - *Economic and Social Cost-Benefit Analyses:* There is a gap in research examining the long-term economic benefits of reducing air pollution, particularly in terms of healthcare savings and improved individual health. Such analyses could assist policymakers in prioritising interventions based on cost-effectiveness and their public health impact [134].
8. *Biomarkers and Early Detection:*
   - *Novel Biomarkers:* Identifying reliable biomarkers that indicate early signs for diseases such as AD, PD, and stroke resulting from long-term air pollution exposure remains a critical research gap. These biomarkers could enable early detection and prevention strategies, potentially reducing the burden of neurological and cardiovascular diseases in polluted environments [135, 136].

6.8 Conclusion

The pathways through which air pollution impacts brain health are intricate and multifaceted, involving a combination of direct neurotoxicity, systemic inflammation, oxidative stress, and vascular dysfunction. Notably, airborne pollutants, especially particulate matter, not only trigger neuroinflammatory responses but also damage neurons and blood vessels, contributing to a wide array of neurological conditions, such as stroke, cognitive decline, and neurodegenerative diseases, like AD and PD. The expanding evidence in the field underscores the urgent need for effective public health interventions to reduce exposure to harmful pollutants and mitigate their impact on brain health. A comprehensive and in-depth understanding of these underlying mechanisms is vital for developing strategies to protect vulnerable populations and prevent the rising burden of air pollution-related neurological conditions globally. Additionally, further investigation into the long-term consequences of air pollution on brain function is crucial to develop preventative measures and advance our knowledge in this intricate field.FundingThis work was supported by the National Health and Medical Research Council (NHMRC) Senior Research Fellowship (1118452) (ARW).

References

1. Peeples, L. (2020). News feature: How air pollution threatens brain health. *Proceedings of the National Academy of Sciences of the United States of America, 117*(25), 13856–13860.
2. Russ, T. C., Reis, S., & van Tongeren, M. (2019). Air pollution and brain health: Defining the research agenda. *Current Opinion in Psychiatry, 32*(2), 97–104.
3. Slomski, A. (2022). Air quality and brain health. *JAMA, 327*(15), 1430–1431.
4. Chen, X., et al. (2020). Respiratory inflammation and short-term ambient air pollution exposures in adult Beijing residents with and without prediabetes: A panel study. *Environmental Health Perspectives, 128*(6), 67004.
5. Taimuri, B., et al. (2022). Air pollution and cerebrovascular disorders with special reference to Asia: An overview. *Annals of Indian Academy of Neurology, 25*(Suppl 1), S3–S8.
6. Rhew, S. H., Kravchenko, J., & Lyerly, H. K. (2021). Exposure to low-dose ambient fine particulate matter PM2.5 and Alzheimer's disease, non-Alzheimer's dementia, and Parkinson's disease in North Carolina. *PLoS One, 16*(7), e0253253.
7. Saha, S., et al. (2021). A perspective on Nrf2 signaling pathway for neuroinflammation: A potential therapeutic target in Alzheimer's and Parkinson's diseases. *Frontiers in Cellular Neuroscience, 15*, 787258.
8. Buccellato, F. R., et al. (2022). The role of glymphatic system in Alzheimer's and Parkinson's disease pathogenesis. *Biomedicine, 10*(9), 2261.
9. Kalecky, K., Ashcraft, P., & Bottiglieri, T. (2022). One-carbon metabolism in Alzheimer's disease and Parkinson's disease brain tissue. *Nutrients, 14*(3), 599.
10. Szeto, J. Y., & Lewis, S. J. (2016). Current treatment options for Alzheimer's disease and Parkinson's disease dementia. *Current Neuropharmacology, 14*(4), 326–338.
11. Cai, M., et al. (2023). Long-term exposure to ambient fine particulate matter chemical composition and in-hospital case fatality among patients with stroke in China. *The Lancet Regional Health Western Pacific, 32*, 100679.

12. Scheers, H., et al. (2015). Long-term exposure to particulate matter air pollution is a risk factor for stroke: Meta-analytical evidence. *Stroke, 46*(11), 3058–3066.
13. Kioumourtzoglou, M. A., et al. (2016). Long-term PM2.5 exposure and neurological hospital admissions in the Northeastern United States. *Environmental Health Perspectives, 124*(1), 23–29.
14. Bhatnagar, A. (2017). Environmental determinants of cardiovascular disease. *Circulation Research, 121*(2), 162–180.
15. Lin, W., et al. (2024). Short-term exposure to air pollution and the incidence and mortality of stroke: A meta-analysis. *The Neurologist, 29*(3), 179–187.
16. Amini, H., et al. (2020). Long-term exposure to air pollution and stroke incidence: A Danish Nurse cohort study. *Environment International, 142*, 105891.
17. Milton, L. A., & White, A. R. (2020). The potential impact of bushfire smoke on brain health. *Neurochemistry International, 139*, 104796.
18. Hahad, O., et al. (2020). Ambient air pollution increases the risk of cerebrovascular and neuropsychiatric disorders through induction of inflammation and oxidative stress. *International Journal of Molecular Sciences, 21*(12), 4306.
19. Khuc, Q. V., et al. (2022). Brain drain out of the blue: Pollution-induced migration in Vietnam. *International Journal of Environmental Research and Public Health, 19*(6), 3645.
20. Schraufnagel, D. E. (2020). The health effects of ultrafine particles. *Experimental & Molecular Medicine, 52*(3), 311–317.
21. Oberdorster, G., et al. (2004). Translocation of inhaled ultrafine particles to the brain. *Inhalation Toxicology, 16*(6–7), 437–445.
22. Cory-Slechta, D. A., & Sobolewski, M. (2023). Neurotoxic effects of air pollution: An urgent public health concern. *Nature Reviews. Neuroscience, 24*(3), 129–130.
23. Jankowska-Kieltyka, M., Roman, A., & Nalepa, I. (2021). The air we breathe: Air pollution as a prevalent proinflammatory stimulus contributing to neurodegeneration. *Frontiers in Cellular Neuroscience, 15*, 647643.
24. Block, M. L., & Calderon-Garciduenas, L. (2009). Air pollution: Mechanisms of neuroinflammation and CNS disease. *Trends in Neurosciences, 32*(9), 506–516.
25. Kwon, H. S., & Koh, S. H. (2020). Neuroinflammation in neurodegenerative disorders: The roles of microglia and astrocytes. *Translational Neurodegeneration, 9*(1), 42.
26. Martikainen, M. V., et al. (2021). TUBE project: Transport-derived ultrafines and the brain effects. *International Journal of Environmental Research and Public Health, 19*(1), 311.
27. Cory-Slechta, D. A., Merrill, A., & Sobolewski, M. (2023). Air pollution-related neurotoxicity across the life span. *Annual Review of Pharmacology and Toxicology, 63*, 143–163.
28. Saveleva, L., et al. (2022). Subacute inhalation of ultrafine particulate matter triggers inflammation without altering amyloid beta load in 5xFAD mice. *Neurotoxicology, 89*, 55–66.
29. Traboulsi, H., et al. (2017). Inhaled pollutants: The molecular scene behind respiratory and systemic diseases associated with ultrafine particulate matter. *International Journal of Molecular Sciences, 18*(2), 243.
30. Lavigne, E., et al. (2019). Spatiotemporal variations in ambient ultrafine particles and the incidence of childhood asthma. *American Journal of Respiratory and Critical Care Medicine, 199*(12), 1487–1495.
31. Guo, C., et al. (2013). Oxidative stress, mitochondrial damage and neurodegenerative diseases. *Neural Regeneration Research, 8*(21), 2003–2014.
32. Bayo Jimenez, M. T., et al. (2023). Noise, air, and heavy metal pollution as risk factors for endothelial dysfunction. *European Cardiology, 18*, e09.
33. Liao, J. K. (2013). Linking endothelial dysfunction with endothelial cell activation. *The Journal of Clinical Investigation, 123*(2), 540–541.
34. Suganya, N., et al. (2016). Reversibility of endothelial dysfunction in diabetes: Role of polyphenols. *The British Journal of Nutrition, 116*(2), 223–246.

35. Liu, F., et al. (2015). Macrophages treated with particulate matter PM2.5 induce selective neurotoxicity through glutaminase-mediated glutamate generation. *Journal of Neurochemistry, 134*(2), 315–326.
36. Pryor, J. T., Cowley, L. O., & Simonds, S. E. (2022). The physiological effects of air pollution: Particulate matter, physiology and disease. *Frontiers in Public Health, 10*, 882569.
37. Lodovici, M., & Bigagli, E. (2011). Oxidative stress and air pollution exposure. *Journal of Toxicology, 2011*, 487074.
38. Zhou, X., et al. (2021). Excess of COVID-19 cases and deaths due to fine particulate matter exposure during the 2020 wildfires in the United States. *Science Advances, 7*(33), eabi8789.
39. Ku, T., et al. (2017). Synergistic effects of particulate matter (PM(2.5)) and sulfur dioxide (SO(2)) on neurodegeneration via the microRNA-mediated regulation of tau phosphorylation. *Toxicology Research (Camb), 6*(1), 7–16.
40. Randolph, A. C., et al. (2019). Blood-brain barrier dysfunction after smoke inhalation injury, with and without skin burn. *Shock, 51*(5), 634–649.
41. D'Angiulli, A. (2018). Severe urban outdoor air pollution and children's structural and functional brain development, from evidence to precautionary strategic action. *Frontiers in Public Health, 6*, 95.
42. Herting, M. M., et al. (2019). Outdoor air pollution and brain structure and function from across childhood to young adulthood: A methodological review of brain MRI studies. *Frontiers in Public Health, 7*, 332.
43. Karceski, S. (2020). Air pollution and brain health. *Neurology, 94*(17), e1872–e1875.
44. Ezkurdia, A., Ramirez, M. J., & Solas, M. (2023). Metabolic syndrome as a risk factor for Alzheimer's disease: A focus on insulin resistance. *International Journal of Molecular Sciences, 24*(5), 4354.
45. Kodavanti, U. P. (2015). Air pollution and insulin resistance: Do all roads lead to Rome? *Diabetes, 64*(3), 712–714.
46. Tham, R., & Schikowski, T. (2021). The role of traffic-related air pollution on neurodegenerative diseases in older people: An epidemiological perspective. *Journal of Alzheimer's Disease, 79*(3), 949–959.
47. Kuntic, M., et al. (2024). Crosstalk between oxidative stress and inflammation caused by noise and air pollution-implications for neurodegenerative diseases. *Antioxidants (Basel), 13*(3), 266.
48. Iaccarino, L., et al. (2021). Association between ambient air pollution and amyloid positron emission tomography positivity in older adults with cognitive impairment. *JAMA Neurology, 78*(2), 197–207.
49. Shi, L., et al. (2021). A national cohort study (2000-2018) of long-term air pollution exposure and incident dementia in older adults in the United States. *Nature Communications, 12*(1), 6754.
50. Krzyzanowski, B., et al. (2024). Air pollution and Parkinson disease in a population-based study. *JAMA Network Open, 7*(9), e2433602.
51. Lind, P. M., & Lind, L. (2020). Are persistent organic pollutants linked to lipid abnormalities, atherosclerosis and cardiovascular disease? A review. *Journal of Lipid and Atherosclerosis, 9*(3), 334–348.
52. Lee, K. K., Miller, M. R., & Shah, A. S. V. (2018). Air pollution and stroke. *Journal of Stroke, 20*(1), 2–11.
53. Jiang, X. Q., Mei, X. D., & Feng, D. (2016). Air pollution and chronic airway diseases: What should people know and do? *Journal of Thoracic Disease, 8*(1), E31–E40.
54. Wolf, K., et al. (2021). Long-term exposure to low-level ambient air pollution and incidence of stroke and coronary heart disease: A pooled analysis of six European cohorts within the ELAPSE project. *Lancet Planetary Health, 5*(9), E620–E632.
55. Stockfelt, L., et al. (2017). Long-term effects of total and source-specific particulate air pollution on incident cardiovascular disease in Gothenburg, Sweden. *Environmental Research, 158*, 61–71.

56. Ma, T., et al. (2022). Long-term air pollution exposure and incident stroke in American older adults: A national cohort study. *Global Epidemiology, 4*, 100073.
57. Fisher, J. A., et al. (2019). Case-crossover analysis of short-term particulate matter exposures and stroke in the health professionals follow-up study. *Environment International, 124*, 153–160.
58. Keller, K., et al. (2023). Air pollution impacts on in-hospital case-fatality rate of ischemic stroke patients. *Thrombosis Research, 225*, 116–125.
59. Newbury, J. B., et al. (2021). Association between air pollution exposure and mental health service use among individuals with first presentations of psychotic and mood disorders: Retrospective cohort study. *The British Journal of Psychiatry, 219*(6), 678–685.
60. Shen, W. T., et al. (2021). Population health effects of air pollution: Fresh evidence from China health and retirement longitudinal survey. *Frontiers in Public Health, 9*, 779552.
61. Zhang, L., et al. (2022). Green space, air pollution, weather, and cognitive function in middle and old age in China. *Frontiers in Public Health, 10*, 871104.
62. Nordin, S., et al. (2023). Impact of comorbidity on symptomatology in various types of environmental intolerance in a general Swedish and Finnish adult population. *Environmental Research, 229*, 115945.
63. McLachlan, J., et al. (2023). Long-term exposure to air pollution and cognitive function in older adults: A systematic review and meta-analysis. *Frontiers in Environmental Health, 2*, 1205443.
64. Bai, K. J., et al. (2021). Exposure to PM(2.5) is associated with malignant pleural effusion in lung cancer patients. *Ecotoxicology and Environmental Safety, 208*, 111618.
65. Perez-Carrascosa, F. M., et al. (2021). Historical exposure to persistent organic pollutants and cardiovascular disease: A 15-year longitudinal analysis focused on pharmaceutical consumption in primary care. *Environment International, 156*, 106734.
66. Tsai, C. Y., et al. (2023). Impact of lifetime air pollution exposure patterns on the risk of chronic disease. *Environmental Research, 229*, 115957.
67. Crous-Bou, M., et al. (2020). Impact of urban environmental exposures on cognitive performance and brain structure of healthy individuals at risk for Alzheimer's dementia. *Environment International, 138*, 105546.
68. To, T, et al. (2015). Health risk of air pollution on people living with major chronic diseases: A Canadian population-based study. *BMJ Open, 5*(9), e009075.
69. Chen, P. C., et al. (2022). A cohort study evaluating the risk of stroke associated with long-term exposure to ambient fine particulate matter in Taiwan. *Environmental Health, 21*(1), 43.
70. Organisation, W.H. Ageing. [Website] 2024 [cited 2024 24/10/2024]. Available from https://www.who.int/health-topics/ageing#tab=tab_1
71. Jones, C. H., & Dolsten, M. (2024). Healthcare on the brink: Navigating the challenges of an aging society in the United States. *NPJ Aging, 10*(1), 22.
72. Powers, W. J., et al. (2019). Guidelines for the early management of patients with acute ischemic stroke: 2019 update to the 2018 guidelines for the early management of acute ischemic stroke: A guideline for healthcare professionals from the American Heart Association/American Stroke Association. *Stroke, 50*(12), e344–e418.
73. Owolabi, M. O., et al. (2021). The state of stroke services across the globe: Report of World Stroke Organization-World Health Organization surveys. *International Journal of Stroke, 16*(8), 889–901.
74. Rochmah, T. N., et al. (2021). Economic burden of stroke disease: A systematic review. *International Journal of Environmental Research and Public Health, 18*(14), 7552.
75. Ma, R., et al. (2023). From crisis to resilience: Strengthening climate action in OECD countries through environmental policy and energy transition. *Environmental Science and Pollution Research International, 30*(54), 115480–115495.
76. Kwilinski, A., Lyulyov, O., & Pimonenko, T. (2024). Reducing transport sector CO2 emissions patterns: Environmental technologies and renewable energy. *Journal of Open Innovation: Technology, Market, and Complexity, 10*(1), 100217.

77. Collazos, J. S. G., Ardila, L. M. C., & Cardona, C. J. F. (2024). Energy transition in sustainable transport: Concepts, policies, and methodologies. *Environmental Science and Pollution Research International, 31*(49), 58669–58686.
78. Blanco, M. N., et al. (2024). Traffic-related air pollution and dementia incidence in the adult changes in thought study. *Environment International, 183*, 108418.
79. McCarron, A., et al. (2025). Piloting co-developed behaviour change interventions to reduce exposure to air pollution and improve self-reported asthma-related health. *Journal of Exposure Science & Environmental Epidemiology, 35*(2), 242–253.
80. Sokhi, R. S., et al. (2022). Advances in air quality research—Current and emerging challenges. *Atmospheric Chemistry and Physics, 22*(7), 4615–4703.
81. Imam, M., et al. (2024). Air quality monitoring using statistical learning models for sustainable environment. *Intelligent Systems with Applications, 22*, 200333.
82. Swamy, G. (2021). Development of an indoor air purification system to improve ventilation and air quality. *Heliyon, 7*(10), e08153.
83. Andre, P., et al. (2024). Globally representative evidence on the actual and perceived support for climate action. *Nature Climate Change, 14*(3), 253–259.
84. Ecker, U. K. H., et al. (2022). The psychological drivers of misinformation belief and its resistance to correction. *Nature Reviews Psychology, 1*(1), 13–29.
85. Cacciatore, M. A. (2021). Misinformation and public opinion of science and health: Approaches, findings, and future directions. *Proceedings of the National Academy of Sciences of the United States of America, 118*(15), e1912437117.
86. Southwell, B. G., et al. (2022). Defining and measuring scientific misinformation. *The Annals of the American Academy of Political and Social Science, 700*(1), 98–111.
87. West, J. D., & Bergstrom, C. T. (2021). Misinformation in and about science. *Proceedings of the National Academy of Sciences of the United States of America, 118*(15), e1912444117.
88. Bhui, K., et al. (2023). Air quality and mental health: Evidence, challenges and future directions. *BJPsych Open, 9*(4), e120.
89. Libby, P., Nahrendorf, M., & Swirski, F. K. (2016). Leukocytes link local and systemic inflammation in ischemic cardiovascular disease: An expanded "cardiovascular continuum". *Journal of the American College of Cardiology, 67*(9), 1091–1103.
90. Brockmeyer, S., & D'Angiulli, A. (2016). How air pollution alters brain development: The role of neuroinflammation. *Translational Neuroscience, 7*(1), 24–30.
91. Henein, M. Y., et al. (2022). The role of inflammation in cardiovascular disease. *International Journal of Molecular Sciences, 23*(21), 12906.
92. Widlansky, M. E., et al. (2003). The clinical implications of endothelial dysfunction. *Journal of the American College of Cardiology, 42*(7), 1149–1160.
93. Drera, A., et al. (2024). Endothelial dysfunction in heart failure: What is its role? *Journal of Clinical Medicine, 13*(9), 2534.
94. Pepin, M. E., & Gupta, R. M. (2024). The role of endothelial cells in atherosclerosis: Insights from genetic association studies. *The American Journal of Pathology, 194*(4), 499–509.
95. Sun, H. J., et al. (2019). Role of endothelial dysfunction in cardiovascular diseases: The link between inflammation and hydrogen sulfide. *Frontiers in Pharmacology, 10*, 1568.
96. Takata, F., et al. (2021). Blood-brain barrier dysfunction amplifies the development of neuroinflammation: Understanding of cellular events in brain microvascular endothelial cells for prevention and treatment of BBB dysfunction. *Frontiers in Cellular Neuroscience, 15*, 661838.
97. Taylor-Clark, T. E. (2020). Air pollution-induced autonomic modulation. *Physiology (Bethesda), 35*(6), 363–374.
98. Basith, S., et al. (2022). The impact of fine particulate matter 2.5 on the cardiovascular system: A review of the invisible killer. *Nanomaterials (Basel), 12*(15), 2656.
99. Shi, L. H., et al. (2020). Long-term effects of PM on neurological disorders in the American Medicare population: A longitudinal cohort study. *Lancet Planetary Health, 4*(12), E557–E565.

100. Yan, W., et al. (2016). NO2 inhalation promotes Alzheimer's disease-like progression: Cyclooxygenase-2-derived prostaglandin E2 modulation and monoacylglycerol lipase inhibition-targeted medication. *Scientific Reports, 6*, 22429.
101. Chang, K. H., et al. (2014). Increased risk of dementia in patients exposed to nitrogen dioxide and carbon monoxide: A population-based retrospective cohort study. *PLoS One, 9*(8), e103078.
102. Sun, S., et al. (2019). Short-term exposure to air pollution and incidence of stroke in the Women's Health Initiative. *Environment International, 132*, 105065.
103. Goldberg, H. C. A. M. S. (2009). The effects of outdoor air pollution on chronic illnesses. *McGill Journal of Medicine, 12*(1), 58–64.
104. Li, L., et al. (2022). Short-term exposure to nitrogen dioxide and ischemic stroke incidence in Shenzhen, China: Modification effects by season and temperature. *Ecotoxicology and Environmental Safety, 239*, 113644.
105. Seposo, X., et al. (2023). Role of oxides of nitrogen in the ozone-cardiorespiratory visit association. *Environmental Pollution, 317*, 120802.
106. Kodavanti, P. R. (2006). Neurotoxicity of persistent organic pollutants: Possible mode(s) of action and further considerations. *Dose-Response, 3*(3), 273–305.
107. Yang, D., et al. (2024). Persistent organic pollutants released from decomposed adipose tissue affect mitochondrial enzyme function in the brain and eyes other than the liver. *Environmental Science and Pollution Research International, 31*(7), 10648–10660.
108. Wang, J., et al. (2018). Penetration of PFASs across the blood cerebrospinal fluid barrier and its determinants in humans. *Environmental Science & Technology, 52*(22), 13553–13561.
109. Shi, L., et al. (2023). Incident dementia and long-term exposure to constituents of fine particle air pollution: A national cohort study in the United States. *Proceedings of the National Academy of Sciences of the United States of America, 120*(1), e2211282119.
110. Haddad, P., et al. (2023). Long-term exposure to traffic-related air pollution and stroke: A systematic review and meta-analysis. *International Journal of Hygiene and Environmental Health, 247*, 114079.
111. Vivanco-Hidalgo, R. M., et al. (2018). Short-term exposure to traffic-related air pollution and ischemic stroke onset in Barcelona, Spain. *Environmental Research, 162*, 160–165.
112. Ljungman, P. L. S., et al. (2018). Long- and short-term air pollution exposure and measures of arterial stiffness in the Framingham heart study. *Environment International, 121*(Pt 1), 139–147.
113. Sunyer, J. (2008). The neurological effects of air pollution in children. *The European Respiratory Journal, 32*(3), 535–537.
114. Rani, P., & Dhok, A. (2023). Effects of pollution on pregnancy and infants. *Cureus, 15*(1), e33906.
115. Calderon-Garciduenas, L., et al. (2014). Air pollution and detrimental effects on children's brain. The need for a multidisciplinary approach to the issue complexity and challenges. *Frontiers in Human Neuroscience, 8*, 613.
116. Costa, L. G., et al. (2019). Developmental impact of air pollution on brain function. *Neurochemistry International, 131*, 104580.
117. de Prado Bert, P., et al. (2018). The effects of air pollution on the brain: A review of studies interfacing environmental epidemiology and neuroimaging. *Current Environmental Health Reports, 5*(3), 351–364.
118. Du, Y., et al. (2016). Air particulate matter and cardiovascular disease: The epidemiological, biomedical and clinical evidence. *Journal of Thoracic Disease, 8*(1), E8–E19.
119. Cserbik, D., et al. (2020). Fine particulate matter exposure during childhood relates to hemispheric-specific differences in brain structure. *Environment International, 143*, 105933.
120. Sun, X., & Li, X. (2023). Editorial: Aging and chronic disease: Public health challenge and education reform. *Frontiers in Public Health, 11*, 1175898.
121. Wu, Y. C., et al. (2015). Association between air pollutants and dementia risk in the elderly. *Alzheimers Dement (Amst), 1*(2), 220–228.

122. Chen, F., et al. (2024). Breathing in danger: Understanding the multifaceted impact of air pollution on health impacts. *Ecotoxicology and Environmental Safety, 280*, 116532.
123. Downward, G. S., et al. (2018). Long-term exposure to ultrafine particles and incidence of cardiovascular and cerebrovascular disease in a prospective study of a Dutch cohort. *Environmental Health Perspectives, 126*(12), 127007.
124. Niu, Y., et al. (2022). Long-term exposure to ozone and cardiovascular mortality in China: A nationwide cohort study. *Lancet Planetary Health, 6*(6), E496–E503.
125. Nobile, F., et al. (2023). Long-term exposure to air pollution and incidence of mental disorders. A large longitudinal cohort study of adults within an urban area. *Environment International, 181*, 108302.
126. Schillemans, T., et al. (2024). OMICs signatures linking persistent organic pollutants to cardiovascular disease in the Swedish mammography cohort. *Environmental Science & Technology, 58*(2), 1036–1047.
127. Urbano, T., et al. (2023). Particulate matter exposure from motorized traffic and risk of conversion from mild cognitive impairment to dementia: An Italian prospective cohort study. *Environmental Research, 222*, 115425.
128. Finch, J., & Conklin, D. J. (2016). Air pollution-induced vascular dysfunction: Potential role of endothelin-1 (ET-1) system. *Cardiovascular Toxicology, 16*(3), 260–275.
129. Mukherjee, S., et al. (2021). Air pollution-induced epigenetic changes: Disease development and a possible link with hypersensitivity pneumonitis. *Environmental Science and Pollution Research International, 28*(40), 55981–56002.
130. Niu, Z., et al. (2021). Association between exposure to ambient air pollution and hospital admission, incidence, and mortality of stroke: An updated systematic review and meta-analysis of more than 23 million participants. *Environmental Health and Preventive Medicine, 26*(1), 15.
131. Ryan, J., & Owen, A. J. (2021). The impact of the climate crisis on brain health. *Nature Aging, 1*(5), 416–417.
132. Lohmann, P. M., et al. (2023). Air pollution and anti-social behaviour: Evidence from a randomised lab-in-the-field experiment. *Social Science & Medicine, 320*, 115617.
133. Nair, A. N., et al. (2022). A review of strategies and their effectiveness in reducing indoor airborne transmission and improving indoor air quality. *Environmental Research, 213*, 113579.
134. Xing, G., Zhang, Y., & Guo, J. (2023). Environmental regulation in evolution and governance strategies. *International Journal of Environmental Research and Public Health, 20*(6), 4906.
135. Ji, H., & Khurana Hershey, G. K. (2012). Genetic and epigenetic influence on the response to environmental particulate matter. *The Journal of Allergy and Clinical Immunology, 129*(1), 33–41.
136. Rider, C. F., & Carlsten, C. (2019). Air pollution and DNA methylation: Effects of exposure in humans. *Clinical Epigenetics, 11*(1), 131.

Chapter 7
Air Pollution and Dementia: Evidence from Epidemiological Studies

Anna Oudin

Keywords Air pollution · Dementia · Alzheimer's disease · Vascular dementia · Cognitive decline · Environmental epidemiology

Abbreviations

AD	Alzheimer's Disease
Aβ	Amyloid-beta
CVD	Cardiovascular disease
CI	Confidence Interval
CRFs	Concentration-response functions
LMICs	Low- and middle-income countries
NFTs	Neurofibrillary tangles
NO_x	Nitrogen Oxides
NO_2	Nitrogen dioxide
OR	Odds Ratio
$PM_{2.5}$	Particulate Matter with an aerodynamic diameter ≤2.5 μm
PM_{10}	Particulate Matter with an aerodynamic diameter ≤10 μm
PAF	Population attributable fraction
RR	Relative Risk
VaD	Vascular Dementia
WHO	World Health Organization

A. Oudin (✉)
Planetary Health Research Group, Division of Occupational and Environmental Medicine, Lund University, Lund, Sweden

Sustainable Health, Department of Epidemiology and Global Health, Umeå University, Umeå, Sweden
e-mail: Anna.oudin@umu.se; Anna.Oudin@med.lu.se

K. M. Kanninen, A. R. White (eds.), *Air Pollution and the Brain*,
https://doi.org/10.1007/978-3-031-99302-2_7

7.1 Introduction

Exposure to ambient (outdoor) air pollution poses significant risks to human health, contributing to a range of serious conditions such as cardiovascular disease, respiratory disorders, and cancer. The World Health Organization estimates that 4.2 million deaths globally were linked to ambient air pollution in 2019 alone (Ambient (outdoor) air pollution). Beyond these well-known impacts, growing research over recent decades highlights the potential harm of ambient air pollution poses to brain health. Evidence that ambient air pollution is associated with psychiatric disorders, cognitive decline, and increased risk of dementia is constantly growing. Estimates vary, but particulate air pollution was suggested to be a contributing cause to 3% of all dementia cases worldwide in the latest "Dementia prevention, intervention, and care: 2024 report of the Lancet standing Commission" to date [1]. In another major study, it was suggested that in 2015, ambient fine air pollution was linked to 2.1 million new cases of dementia and 600,000 deaths worldwide, resulting in a total loss of 7.3 million disability-adjusted life years [2]. This highlights the potential significant public health impact of air pollution on cognitive health and overall well-being.

These numbers are expected to rise as urbanization increases. As of 2024, approximately 4.5 billion people, or about 56% of the global population, reside in urban areas. This figure reflects a continuing trend of urbanization, with projections indicating that the urban population will continue to grow in the coming years. Climate change is furthermore expected to worsen air quality by increasing ozone and particulate matter levels, particularly through events like wildfires. Additionally, the global population is aging, further emphasizing the need to understand how air pollution affects dementia, as more individuals live longer in polluted urban environments.

In this chapter, we aim to explore and clarify the epidemiological evidence linking air pollution to dementia. This includes a brief survey of the earliest epidemiological studies, but mostly a review of reviews. We particularly focus on different pollutants, consider whether different subtypes of dementia exhibit varying vulnerabilities to ambient air pollution, and identify existing knowledge gaps. Additionally, we will discuss the timing of exposure, examining critical windows when individuals might be most susceptible, and investigate other factors that could mediate or enhance susceptibility to the effects of ambient air pollution. We focus exclusively on ambient air pollution, which will be hereafter referred to as "air pollution" in this context and will not specifically include or discuss indoor or household air pollution.

7.2 Dementia

Dementia is a chronic, progressive condition affecting cognitive functions such as memory, language, and executive functioning, often leading to a significant decline in daily living abilities. According to the WHO, dementia is an umbrella term for multiple diseases, with age being the most significant risk factor. However, it is not considered a normal part of aging.

The American Psychiatric Association's DSM-IV defined dementia as a decline in cognitive abilities, with common features across all subtypes including memory deficits and additional impairments such as aphasia (speech difficulties), apraxia (motor dysfunction), agnosia (object recognition problems), or deficits in executive functioning. In DSM-V (2013), the term "major neurocognitive disorder" replaced "dementia," but in this chapter, we still use the word "dementia".

As of 2023, approximately 55 million people living with dementia worldwide, and this number is projected to rise to around 78 million by 2030 and 139 million by 2050. The societal costs associated with dementia are substantial. In 2019, the total economic impact of dementia was estimated at $1.3 trillion, with informal caregivers shouldering about half of these costs. This highlights not only the financial strain but also the emotional toll on families and caregivers.

Various risk factors for dementia have been identified, including non-modifiable ones like gender and genetic predispositions, particularly the APOE-ε4 gene. The 2020 report by the Lancet Commission on dementia prevention, intervention, and care highlighted air pollution as an emerging modifiable risk factor for dementia [3]. This addition complements other established factors such as low education, hypertension, smoking, obesity, hearing loss, physical inactivity, depression, diabetes, social isolation, excessive alcohol use, and traumatic brain injury. Two new risk factors were added in the Lancet report from 2024, namely vision loss and high cholesterol [1]. The Lancet reports underscore that addressing these risk factors through preventive measures could significantly delay or even prevent dementia onset, particularly in vulnerable populations. Given the global burden of dementia, even modest reductions in incidence could yield substantial benefits, both for individuals and society. It is also interesting to note that the population attributable fraction of dementia cases attributed to air pollution raised from 2.3% in the report from 2020 [3] to 3% in the report from 2024 [1].

Dementia subtypes are typically divided into primary (neurological diseases), vascular (due to cardiovascular issues), and secondary (from trauma or infections). AD is characterized by memory loss, hallucinations, and behavioural changes, while VaD, the second most common subtype, results from restricted blood flow to the brain due to cardiovascular problems, often leading to attention and planning deficits. Mixed dementia, where a person presents both AD and VaD pathologies, is increasingly recognized as a significant category. Research suggests that many cases exist on a spectrum between AD and VaD, rather than being purely one or the other.

7.2.1 Alzheimer's Disease

AD is the most common form of dementia, accounting for 60–80% of cases [4]. It is marked by progressive neuronal death, leading to a decline in cognitive functions, particularly memory and language [5]. AD also manifests in psychiatric symptoms, such as hallucinations and delusions, along with behavioural changes like increased agitation and aggression [6].

Prevalence varies by region, with higher rates in industrialized nations due to greater life expectancy. Age is the most significant risk factor, with risk beginning to rise notably after age 60. Studies suggest that 5–9% of individuals over 60 are affected by AD, and prevalence can be as high as 48% for men and 75% for women in those over 90 [5, 7].

AD involves two main culprits at the cellular level: amyloid-beta (Aβ) plaques and neurofibrillary tangles (NFTs) [8]. Aβ plaques are clumps of protein fragments that build up outside neurons and were once thought to be the primary cause of the disease, as suggested by the "amyloid cascade hypothesis." According to this idea, these plaques interfere with neuron communication, leading to AD. But the story turned out to be more complicated—some people with large amounts of Aβ plaques don't show significant symptoms, while others with AD may not have as much plaque buildup [9].

The second hallmark of AD is the accumulation of tau proteins inside neurons, which form NFTs [9, 10]. Normally, tau proteins help stabilize the internal structure of neurons. However, when tau proteins become damaged, they clump together into tangles, disrupting the transport system within neurons. This process causes neuronal dysfunction and is closely tied to cognitive decline in AD patients. Interestingly, the progression of tau tangles is often a better predictor of cognitive decline than Aβ plaque buildup, making tau pathology a critical aspect of understanding AD [11].

7.2.2 *Vascular Dementia*

VaD accounts for about 15–20% of all dementia cases [12]. VaD arises due to cardiovascular problems that impair blood flow to the brain, leading to neuronal death from oxygen deprivation [13]. As with other dementias, advancing age is a significant risk factor for VaD, but VaD is also closely linked to cardiovascular disease (CVD). Consequently, VaD has a higher mortality rate than AD due to its association with heart-related conditions. Unlike AD, VaD is more common in men, likely because men have a higher risk of stroke, a major cause of VaD [14].

VaD exhibits more variation in symptoms compared to AD, but common features include cognitive deficits, especially those related to the frontal lobe, which result in impairments in attention, planning, and processing speed [15]. Although major cardiovascular events like stroke are strong risk factors for VaD, most cases arise from smaller vascular lesions caused by subcortical ischemic vascular disease, often linked to atherosclerosis or diabetes [16]. Additionally, damaged blood vessels disrupt trophic signalling, which is essential for the maintenance and repair of neurons [17].

7.2.3 Mixed Dementia

Mixed dementia refers to cases where multiple underlying causes contribute to the development of the disease, most commonly a combination of AD and VaD. It is estimated that 20–40% of dementia cases fall into this mixed category [18, 19]. AD and VaD share several risk factors, and there is evidence of synergistic effects, where symptoms of one condition can exacerbate or accelerate the other. Research suggests that "pure" AD or VaD cases are less common than mixed forms, with most dementia patients showing features of both conditions [19]. This has led to an emerging view of dementia as existing on a spectrum between AD and VaD, rather than being classified as strictly one or the other.

The differences between Alzheimer's disease and vascular dementia can generally be categorized into four main types [20, 21]:

1. *Pathophysiology Differences:*

   - *Alzheimer's Disease:* The key mechanisms involve amyloid-beta deposition, tau protein abnormalities, and progressive neuronal loss. Genetic factors, such as carrying the APOE-ε4 allele, and lifestyle factors like poor diet and low cognitive engagement, are more strongly associated with AD.
   - *Vascular Dementia:* VaD is caused by conditions that disrupt the blood supply to the brain, leading to ischemic lesions and white matter damage. Risk factors like hypertension, diabetes, stroke, and other cardiovascular diseases are central to VaD pathology.

2. *Vascular vs. Neurodegenerative Risk:*

   - *Cardiovascular Risk Factors* (e.g., hypertension, high cholesterol, diabetes, smoking) have a stronger association with VaD because of the direct impact on blood vessels. Studies suggest that while these factors also increase AD risk, they are more predictive of VaD due to the role of cerebral blood flow in its development.
   - *Neurodegenerative Risk Factors* such as genetics (e.g., APOE-ε4), low physical activity, and cognitive inactivity are more strongly linked with AD, as they directly influence brain degeneration rather than vascular health.

3. *Age of Onset:*

   - Both dementias share age as a major risk factor, but the typical onset and progression differ. AD tends to present with earlier cognitive decline (often starting in the 60s and 70s), while VaD may occur following a stroke or cardiovascular event, making its onset more variable depending on the timing of these vascular events.

4. *Differential Impact of Lifestyle:*

   - *Exercise and Diet* can affect both AD and VaD, but the pathways differ. Regular exercise and a Mediterranean diet may help reduce the risk of AD by improving brain plasticity and reducing amyloid accumulation, while for VaD, they help primarily by reducing cardiovascular risk.

7.3 Air Pollution

Air pollution refers to the presence of harmful substances in the air, which can include particulate matter (PM), gases like nitrogen dioxide (NO_2) and ozone, and chemicals from industrial processes, vehicles, and household activities. These pollutants can harm both the environment and human health. Major sources include traffic emissions, power plants, agriculture, and domestic fuel burning.

Globally, air pollution is one of the leading environmental risk factors for disease. According to the World Health Organization (WHO), it contributes to around 7 million premature deaths each year, particularly from heart disease, stroke, chronic obstructive pulmonary disease, lung cancer, and respiratory infections [22–24]. In addition to respiratory and cardiovascular diseases, evidence is growing that long-term exposure to air pollution may also increase the risk of neurodegenerative conditions such as dementia, as the focus of this chapter. Populations in low- and middle-income countries bear the heaviest burden, with limited access to clean air and high exposure to pollutants [22]. Efforts to reduce air pollution, such as implementing cleaner energy sources, improving transportation systems, and enforcing stricter environmental regulations, can have a profound impact on public health, reducing the global health burden associated with polluted air [25].

Currently, PM with an aerodynamic diameter ≤10 μm (PM_{10}) and ≤2.5 μm ($PM_{2.5}$) are the most widely used metrics for studying air pollution and its health impacts, and there is substantial evidence linking especially $PM_{2.5}$ ("fine" PM) to negative health outcomes. It is typically measured as a density: $\mu g/m^3$. In some studies, other markers like nitrogen oxides (NO_x) may provide a better indication of harmful pollution, depending on the quality of the data, models used, and the specific pollution sources in the area under study. It is generally accepted that particulate matter, especially fine and ultrafine particles, poses a greater health risk than NO_2; however, although there seem to be independent effects by NO_2, albeit smaller [26]. Multipollutant studies are becoming more common, allowing researchers to differentiate the health impacts of pollutants like $PM_{2.5}$ from those of NO_2. However, the conclusions from these studies depend heavily on the accuracy of the measurements used for each pollutant. The challenge lies in the complexity of air pollution, as it consists of numerous components, many of which have not been thoroughly investigated in health studies. As a result, the health effects of some pollutants remain unclear or entirely unknown. Expanding research to include less-studied elements of the air pollution mix is crucial to deepen our understanding of their potential health risks. Future research will likely help identify the most harmful components of air pollution for different health outcomes.

7.4 Air Pollution and Dementia: Mechanisms

Several pathways have been proposed to explain how air pollutants may affect the brain (see Chaps. 2, 3 and 4). In short, the two direct routes include: (1) the olfactory route, where inhaled particles, particularly PM, enter neurons in the nasal cavity and travel to the brain via the olfactory bulb—an effect observed in rodents and supported by human studies showing increased olfactory deterioration in highly polluted areas; and (2) the blood-brain barrier route, where particles that enter the bloodstream from the lungs may cross into the brain [27–35]. Additionally, systemic inflammation in other organs, like the lungs, may trigger immune responses in the brain. Air pollution seems to play a significant role in triggering immune responses that may contribute to the development of dementia [36]. When PM enters the brain, it can initiate chronic inflammation, impairing the brain's immune cells, such as microglia [37]. Normally responsible for clearing harmful proteins like amyloid-beta plaques, chronic activation of microglia due to pollutants can reduce their effectiveness and worsen neurodegeneration.

This prolonged inflammatory state accelerates harmful processes such as tau phosphorylation, which leads to neurofibrillary tangles and exacerbates cognitive decline. Additionally, oxidative stress caused by PM exposure disrupts the balance between antioxidants and reactive oxygen species, further driving neuronal damage. The emerging field of immunopsychiatry explores how these immune responses influence brain health, offering new insights into the mechanisms by which air pollution may contribute to dementia and suggesting potential therapeutic interventions that target immune pathways.

The risk induced by air pollution for AD and VaD is likely to differ due to the distinct pathological mechanisms underlying these two forms of dementia. AD is primarily characterized by the accumulation of amyloid-beta plaques and neurofibrillary tangles in the brain, which lead to the loss of neuronal function and cognitive decline. In contrast, VaD results from reduced blood flow to the brain due to cardiovascular problems such as stroke, atherosclerosis, or other vascular conditions, which cause neuron damage through oxygen deprivation.

While there is overlap in risk factors, such as age and certain lifestyle factors, the differences in underlying pathology between AD and VaD make it likely that the specific risk factors for each will vary, such as air pollution. AD is more influenced by neurodegenerative processes and genetic predispositions, whereas VaD is more strongly linked to cardiovascular health and vascular risk factors. This distinction highlights the need for stratification by subtype when studying air pollution and dementia, when possible [38].

When discussing air pollution as a risk factor for dementia, it is furthermore important to discuss depression since dementia and depression are closely linked [39]. Depression can be both an early symptom and a potential risk factor for dementia. Individuals with a history of depression, particularly in midlife, appear to have a higher risk of developing dementia later in life. This relationship may be due to several mechanisms, such as chronic inflammation, vascular changes, and the impact of depression on stress hormones, all of which can affect brain health. Depression can accelerate cognitive decline, worsen memory, and contribute to other neurodegenerative changes. Some studies even suggest that depression may trigger changes in the brain that predispose individuals to dementia, such as hippocampal atrophy or impaired neuroplasticity. On the other hand, dementia itself can lead to depression, as individuals with cognitive decline struggle with the emotional toll of losing their independence and memory. Given that air pollution is an emerging risk factor for depression [40] it is crucial to understand the relationship between these conditions and including the role of air pollution.

Another important factor to discuss is that the APOE gene, a significant risk factor for dementia, is responsible for encoding apolipoprotein E, a protein involved in cholesterol transport and metabolism throughout the body, including the central nervous system (CNS) [41]. Within the CNS, apolipoprotein E is primarily produced by glial cells like astrocytes and released into the cerebrospinal fluid. Its functions include supporting neuronal health, promoting anti-inflammatory responses, and facilitating the clearance of amyloid-beta plaques. The APOE gene exists in three alleles—ε2, ε3, and ε4—with the ε4 allele being a well-established genetic risk factor for dementia, particularly AD [42]. Individuals with one ε4 allele (heterozygotes) have a threefold increased risk of late-onset AD, while those with two ε4 alleles (homozygotes) experience an eightfold increase in risk. Additionally, the presence of APOE-ε4 is linked to an earlier onset of AD, often by 10–20 years [43]. APOE-ε4 is also associated with VaD and cardiovascular diseases (CVD), both of which increase dementia risk. It has been linked to elevated levels of low-density lipoprotein cholesterol, a major risk factor for CVD, and to degeneration of blood vessels, including capillaries in the hippocampus, a key brain region affected in dementia. This dual association with both AD and VaD underscores APOE-ε4's broad influence on dementia-related pathologies.

7.5 Review of Reviews

Air pollution as a risk factor for developing dementia was first reported in epidemiological literature in 2014. There was a series of studies conducted in Taiwan that demonstrated that dementia incidence is significantly associated with exposure to various air pollutants, including carbon monoxide, NO_2, ozone, and $PM_{2.5}$ [44], as well as PM_{10} [45]. In Scandinavia, Oudin et al. observed associations between individual-level NO_x exposure (as an indicator for traffic-related air pollution) and increased dementia risk while adjusting for individual-level risk factors, including cardiovascular disease [46]. In that study, the cases of dementia were validated, and AD and VaD were analysed separately, with no clear differences in association. One notable study published in 2017 by Chen et al. followed over two million residents aged 55 and older in Ontario, Canada, over a 12-year period [47, 48]. This research revealed that both proximity to major roads and exposure to $PM_{2.5}$ and NO_x were associated with an increased risk of dementia, with air pollution attributed to 6% of dementia cases.

The first systematic review of the literature was published in 2016 by Power and colleagues [49]. At the time, they found 18 studies on air pollution and cognitive outcomes, and they found that nearly all reported an adverse association between at least one pollutant and a dementia-related outcome. The authors stated that few studies focused on within-person cognitive or pathological changes, which offer stronger causal evidence. Despite potential biases, such as misclassification of dementia cases from health records and reliance on recent air pollution exposure as a proxy for long-term exposure, differential selection biases likely skew results toward protective associations, the authors concluded that the observed adverse effects are likely valid. Unmeasured confounding was also deemed unlikely to explain these associations. The review identified key challenges, including the underdiagnosis of dementia in the community and varying levels of exposure misclassification across pollutants, making it difficult to determine which pollutants are most causally linked to dementia. Overall, the authors concluded that evidence at the time supported a connection between air pollution and dementia, but future, more refined designs and analyses are needed to close evidentiary gaps and guide potential interventions [50].

Another early review included risk factors into six groups: air quality, toxic heavy metals, other metals, trace elements, occupational exposures, and miscellaneous environmental factors [51]. The review noted that few studies adopted a life course approach. At least moderate evidence was found linking several risk factors to cognitive impairment, including air pollution, aluminum, silicon, selenium, pesticides, vitamin D deficiency, and exposure to electric and magnetic fields at the time [51].

In 2018, a review on air pollution related to both stroke and dementia was published, with a focus on both short-term and long-term effects [52]. The authors concluded that, despite quite limited evidence at the time, new pathophysiological mechanisms linking air pollution to cerebrovascular and neurodegenerative diseases highlighted the importance of considering air pollution as a modifiable risk factor for stroke and dementia [52].

In 2019, a systematic review was published that explored the link between air pollution and cognitive decline or dementia [38]. In conclusion, the review suggested growing evidence that air pollution, particularly fine particulate matter and nitrogen oxides, is associated with an increased risk of dementia. Variation in statistical methods and use of quantiles for pollutant exposure prevented the authors from conducting a meta-analysis [38].

Another review from 2019 highlights current research findings while addressing the methodological challenges and opportunities in this field, and focuses not only on air pollution but also on noise [53]. The authors conclude that although the evidence remains limited, particularly regarding chronic noise exposure, high levels of these pollutants have been linked to accelerated cognitive decline, potentially through cerebrovascular events or the development of Alzheimer's disease [53].

In another review, from 2020, the potential associations between air pollution, noise, and the lack of green spaces as contributors to AD are discussed [54]. The review states that although epidemiological studies provide substantial support for a link between air pollution and AD, most epidemiological research tends to overlook a multi-exposure approach that could clarify these relationships. It is furthermore concluded that evidence regarding noise and lack of greenness is more limited than for air pollution as risk factors for dementia (AD in this review). The author further emphasizes the need for future studies to adopt a multi-exposure approach to effectively disentangle the effects of air pollution, noise, and green space availability [54].

In a study published in 2020, the arguably first meta-analysis was published to examine the relationship between air pollution exposure and the risk of cognitive impairment (including dementia) based on longitudinal cohort studies [55]. The pooled Relative Risk (RR) of cognitive impairment increased by 1.08 (95% Confidence Interval, CI = 1.03, 1.13) for every 5 μg/m^3 increment in $PM_{2.5}$. No significant associations were observed between NO_2/NOx or ozone and cognitive impairment. In subgroup analyses, the significant positive association for $PM_{2.5}$ exposure was particularly evident in studies involving population studies conducted in North America and those with follow-up durations exceeding 10 years [55].

In 2020, in the "Dementia prevention, intervention, and care: 2020 report of the Lancet Commission" air pollution it was stated that new evidence supports the addition of three modifiable risk factors—excessive alcohol consumption, head injury, and air pollution—to the 2017 Lancet Commission's life-course model for dementia prevention, intervention, and care [3, 56]. Notably, the population attributable fraction (PAF) for air pollution in people aged 65 years and older was 2.3%, which is higher than, e.g. that for diabetes (1.1%) and physical inactivity (1.6%). Although the review did not focus specifically on air pollution or discuss its effects on sensitive populations, dementia subtypes, or the shape of the dose-response relationship, it did hypothesize that air pollution might influence dementia through vascular mechanisms and cited Peters et al. for that [38]. The review also highlighted that dementia is rising faster in low- and middle-income countries (LMICs) than in high-income countries, driven by aging populations and a higher prevalence of potentially modifiable risk factors. Consequently, preventive interventions could yield the greatest impact in reducing dementia cases in LMICs. This is very relevant in an air pollution perspective since the air pollution levels are often higher in LMICs than in high-income countries. In the updated version of the report from 2024, the authors estimate the PAF to 3% [1], as compared to 2.3% in the report of 2020 [3]. The authors furthermore highlight that as the body of evidence linking air pollution and dementia expands, it would be valuable to standardize study designs, reporting practices, and analytical approaches [1]. They argue that such consistency would enable clearer comparisons across studies and support a deeper, more detailed understanding of the relationship between specific air pollutants and dementia risk. They further state that due to the strong connections between socioeconomic factors, household conditions, and pollution exposure, effectively controlling for residual confounding in these studies presents a complex challenge [1].

In 2020, a systematic review and meta-analysis were conducted to assess the associations between air pollution and the incidence of AD [57]. The researchers included publications up to March 2020. The meta-analysis included nine studies, revealing that the Odds Ratio (OR) per 10 μg/m^3 increase in $PM_{2.5}$ was 1.95 (95% CI: 0.88–4.30). For other pollutants, the OR was 1.03 for Ozone (95% CI: 0.68–1.57), 1.00 for NO_2 (95% CI: 0.89–1.13), and 0.95 for PM_{10} (95% CI: 0.91–0.99). The overall OR for the five air pollutants related to AD was 1.32 (95% CI: 1.09–1.61), indicating a positive association between ambient air pollution and the incidence of AD. Subgroup analysis indicated that the OR in heavily polluted regions was significantly higher at 2.20 compared to 1.06 in lightly polluted regions [57].

In a review published in 2021, Delgado-Saborit and colleagues assessed epidemiological evidence on air pollution's effects on cognitive function, decline, and dementia, following the Bradford Hill criteria for causality [58]. The authors concluded that evidence consistently showed that chronic exposure to air pollution was associated with reduced global cognition and impairments in specific cognitive domains, such as visuospatial abilities. The authors furthermore state that cognitive decline and increased dementia risk have frequently been linked to air pollution exposure. Neuroimaging studies have found associations between air pollution and reductions in white matter volume, gray matter, and corpus callosum size, as well as larger ventricular volumes. However, findings on small vessel disease markers and hippocampal volume were more mixed. Studies on neuroinflammation furthermore also suggested a link with air pollution. The authors furthermore stated that while many studies had adjusted for traditional confounding factors, residual confounding cannot be entirely ruled out. The available evidence at the time met several Bradford Hill criteria for causality, suggesting that air pollution may indeed contribute to cognitive decline and dementia. However, at the time, the diversity of pollutants studied and outcomes measured made the authors conclude that it was difficult to attribute these effects to any single pollutant and prevented a meta-analysis from being appropriate.

In an updated systematic review [50] of the review of Power and colleagues [49] on the relationship between air pollution and dementia, the authors identify a rapidly expanding evidence base since the previous review. The analysis included 66 papers (49 of which were new) with cognitive function remaining the most common outcome examined, and PM the most frequently studied pollutant. The authors state that, notably, recent studies have improved exposure estimation methods and increasingly focused on cognitive changes, neuroimaging, and dementia incidence. However, methodological issues, such as reliance on administrative records to diagnose dementia and the risk of selection bias, persisted. Among the 35 studies that met strict quality standards, those investigating $PM_{2.5}$ generally supported an association with cognitive decline. In contrast, findings regarding PM_{10}, NO_2, and NO_x were inconclusive, and there was insufficient evidence to determine the impact of ozone. Only a few studies met quality criteria for dementia-specific outcomes. In conclusion, the authors stated that at the time, although a consistent adverse link between $PM_{2.5}$ and cognitive decline emerged, the review highlighted the ongoing challenges in the field, particularly regarding methodology, that must be addressed to draw stronger conclusions [50].

A systematic review published in 2021 aimed to summarize various natural, physical, and social environmental factors associated with age-related cognitive impairment and dementia [59]. High-to-moderate quality evidence suggested increased risks associated with exposure to PM2.5 (HR = 1.24, 95% CI: 1.17–1.31), NO_2 (HR = 1.07, 95% CI: 1.02–1.12), aluminium (OR = 1.35, 95% CI: 1.14–1.59), solvents (OR = 1.14, 95% CI: 1.07–1.22), and proximity to roads (OR = 1.08, 95% CI: 1.04–1.12). Conversely, increased social contact (HR = 0.82, 95% CI: 0.76–0.90) and more greenery (OR = 0.97, 95% CI: 0.95–0.995) were found to be protective factors. Low-to-very low-quality evidence suggested risks from electromagnetic fields, pesticides, SO_2, neighbourhood socioeconomic status, and rural living, while more community cultural engagement was seen as potentially protective. No statistically significant associations were noted for PM_{10}, NO_x, noise, silicon, community groups, and temperature. For the remaining 22 factors, only descriptive analyses were possible due to insufficient studies or a lack of information in that study. The review thus highlighted that the strongest evidence for environmental exposures related to age-related cognitive impairment and dementia was seen for air pollution, particularly $PM_{2.5}$ and NO_2 [59].

In a study from 2021, ambient $PM_{2.5}$ pollution was specifically linked to various adverse health effects, including a potential role in high-prevalence diseases like dementia [2]. In response, researchers conducted a systematic review of the epidemiologic literature, focusing on various exposure sources. They estimated that in 2015, ambient $PM_{2.5}$ pollution could be responsible for 2.1 million new dementia cases and 0.6 million deaths globally, resulting in a combined loss of 7.3 million disability-adjusted life years. The study highlighted significant uncertainties in the relationship and called for more epidemiological studies, especially in high-exposure regions [2].

In a study from 2022, where the aim was to analyze the existing evidence on the association between $PM_{2.5}$ and dementia, specifically AD and VaD separately, a systematic review and meta-analysis were done [60]. A total of 20 articles were selected for review, of which 18 reported on the long-term effects of $PM_{2.5}$ on dementia, 13 focused on AD, and eight examined VaD. The pooled HRs revealed that for each 10.0 $\mu g/m^3$ increase in $PM_{2.5}$, the risk of dementia increased by 1.40 (95% CI 1.23, 1.60), AD by 1.47 (95% CI 1.22, 1.78), and VaD by 2.00 (95% CI 1.30, 3.08). The findings suggested that long-term exposure to $PM_{2.5}$ may significantly elevate the risk of developing dementia, including both AD and VaD [60].

A systematic review synthesized all published original studies on the cognitive effects of air pollution across the life course [61]. The findings consistently demonstrate that various air pollutants negatively impact cognitive function throughout different stages of life, including dementia. Despite this, a notable gap exists in research from LMICs, which experience high levels of air pollution and increasing dementia rates. Most studies were concentrated in high- and upper-middle-income countries, leaving a significant knowledge gap regarding how air pollution impacts cognitive health in LMICs.

In a meta-analysis published in 2023, a systematic search of databases identified peer-reviewed epidemiological studies reporting on all-cause dementia, cognitive decline, AD, and VaD linked to exposure to particulate matter ($PM_{2.5}$ and PM_{10}), ozone, NO_2, and NO_x. The analysis, using a random-effects model, showed that for all-cause dementia, $PM_{2.5}$ and NO_2 had pooled HRs of 1.03 and 1.00, respectively, while NO_x had an HR of 1.00. For AD, $PM_{2.5}$ had an HR of 1.08, while ozone showed an HR of 1.02, although the association with ozone was less clear. For VaD, $PM_{2.5}$ was associated with an increased risk, with an HR of 1.03. Results for NO_2 and NO_x were equivocal, and meta-analysis could not be performed for cognitive decline due to variability in assessment methods. Overall, $PM_{2.5}$ exposure emerged as a risk factor for all-cause dementia, AD, and VaD, with ozone potentially being linked to AD.

In a systematic review and meta-analysis published in 2023 by Abolhasani and colleagues, studies on the association between air pollution and dementia incidence were included [62]. Population-based cohort studies that provided HRs for dementia in relation to air pollutants such as $PM_{2.5}$, NO, NO_2, or ozone were included. The analysis focused on adults over 40 years old. A total of 20 studies were included in the systematic review, with 17 providing data for meta-analysis. The combined population of these studies was 91,391,296, of which 6% (5,521,111) were diagnosed with dementia. The meta-analysis included 12 studies for $PM_{2.5}$, 5 for NO_x, 6 for NO_2, and 4 for ozone. The results showed a 3% increase in dementia risk per 1 μg/m^3 increment of $PM_{2.5}$ (HR: 1.03; 95% CI [1.02–1.05]), although there was high heterogeneity. The associations with NO_x (HR: 1.05; 95% CI [0.99–1.13]), NO_2 (HR: 1.03; 95% CI [1.00–1.07]), and Ozone (HR: 1.01; 95% CI [0.91–1.11]) were less clear, with high variability across studies. Overall, the evidence in that review thus suggested a significant association between $PM_{2.5}$ exposure and dementia incidence, while the associations with NO_x, NO_2, and ozone were less conclusive. The small number of studies and the high degree of heterogeneity highlighted the need for careful interpretation of these results.

In another systematic review and meta-analysis aimed at investigating the role of air pollutants in the risk of dementia, accounting for study-specific factors that might influence outcomes, data were gathered [63]. The review included studies of adults aged 18 or older, with longitudinal follow-up, and considered pollutants based on US Environmental Protection Agency criteria. These studies assessed exposure to ambient air pollution, averaged over a year or more, and examined the association with clinical dementia. 51 studies were included, many of which carried a high risk of bias, though often in ways that skewed results toward the null. Fourteen studies contributed to a meta-analysis for $PM_{2.5}$, finding an overall HR per 2 μg/m^3 of $PM_{2.5}$at 1.04 (95% CI: 0.99–1.09). Studies using active case ascertainment yielded a higher HR of 1.42 (1.00–2.02), while those using passive ascertainment reported an HR of 1.03 (0.98–1.07). Nitrogen dioxide showed an overall HR per 10 μg/m^3 of 1.02 (0.98–1.06) across nine studies, and nitrogen oxide an HR of 1.05 (0.98–1.13) across five studies. Ozone showed no clear association with dementia, with an HR per 5 μg/m^3 of 1.00 (0.98–1.05). The results suggested that $PM_{2.5}$ might be a risk factor for dementia, alongside nitrogen dioxide and nitrogen oxide, although data were more limited for the latter. The hazard ratios were subject to significant limitations, requiring cautious interpretation. Differences in outcome ascertainment and exposure assessment across studies posed challenges, and the authors identified a need for further research, especially focusing on critical exposure periods and pollutants beyond $PM_{2.5}$ [63].

A systematic review and meta-analysis of 15 population-based cohort studies published in 2023, mostly from high-income countries, found inconsistent results regarding AP and dementia risk [25]. The authors suggested that important factors like smoking are lacking in some studies, and suggest depression as a potential confounding factor that should be accounted for, but is not in most studies. A meta-analysis of existing data plus data from two cohorts in China showed a significant association between AP and dementia, with increased relative risks for particulate matter ($PM_{2.5}$. PM_{10}), carbon monoxide, nitrogen dioxide, and nitrogen oxides, but not ozone [25].

It should also be mentioned that in their review of evidence for associations between air pollutants and morbidity, Forastiere and colleagues concluded that the inclusion of concentration–response functions (CRFs) for dementia, alongside other health outcomes in health risk assessments, could enhance health risk assessments of long-term air pollution exposure [23]. While dementia was classified under greater uncertainty than asthma in children, chronic obstructive pulmonary disease, ischemic heart disease events, stroke, hypertension, and lung cancer in relation to $PM_{2.5}$ exposure, the authors conclude that its potential association with air pollution warrants attention. The study emphasized the need for ongoing research to refine these associations [23].

This review of reviews on the link between air pollution and dementia underscores significant concerns about the cognitive impacts of environmental pollutants. Among these, $PM_{2.5}$ consistently shows a stronger association with dementia than other pollutants, though some reviews also report links with PM_{10}, NO_2, NO_x, and ozone. However, many reviews emphasize the need for more research, particularly on pollutants beyond $PM_{2.5}$.

One review highlighted that studies using active case ascertainment reported higher hazard ratios compared to those relying on passive ascertainment, suggesting that validation of diagnoses might capture a stronger association between air pollution and dementia risk. Several reviews furthermore highlighted that most studies do not distinguish between Alzheimer's disease and vascular dementia; however, when analysed separately, associations appear for both subtypes.

Another review observed elevated hazard ratios in areas with high pollution levels compared to regions with lower pollution for the same unit increase in exposure. Conversely, several studies noted that most studies were conducted in areas with relatively low pollution levels, yet the concentration-response function still appeared comparatively steep, indicating a potentially robust effect even at lower levels of exposure. When it comes to mortality, there is evidence for the CRF to be steeper at lower levels of air pollution and in studies where the air pollution models can capture near-source contrasts, which may partly explain discrepancies based on concentration range [64]. Much less information on the shape of the CRF for morbidity than mortality, but this is a very relevant topic to focus on in future studies. This also relates to the lack of research in low- and middle-income countries is also flagged, where air pollution concentrations are often higher than in high-income countries. This is a caution that findings from high-income countries may not necessarily apply to low- and middle-income countries. The absence of multi-exposure studies, including multi-pollutant analyses and those considering both air pollution and noise or greenness, is another gap noted.

Despite these limitations, air pollution remains the most studied environmental pollutant linked to dementia, with the strongest evidence supporting its role. Indications exist that other pollutants and chemicals may elevate dementia risk. Finally, the role of cardiovascular morbidity in the association between air pollution and dementia was discussed, but remains an area for further research. Only a small number of studies met quality standards specific to dementia outcomes, reinforcing the need for careful interpretation of findings.

Below, we discuss certain of the issues raised in separate sub-chapters.

7.6 Timing of Exposure

Relating to the need to investigate critical time windows, the field of air pollution and health has faced criticism, particularly due to the absence of comprehensive, long-term assessments of both exposure and outcomes. This is especially relevant for cognitive health outcomes, where evidence suggests that early life exposure may be of importance. Studies have rarely covered the full life-course of individuals, leaving unanswered questions about the most critical periods during which exposure to air pollution might harm the brain, as discussed by [65]. Many studies, while claiming to investigate long-term exposure, have typically estimated air pollution levels at only a single point in time, often overlooking exposure variability over an individual's lifespan. This gap in understanding is further complicated by the lack of historical air pollution data, as systematic modelling of air pollutants only began a few decades ago. Additionally, there is limited information on participants' residential histories a long way back in time, making it difficult to track exposure across different life stages.

Despite the expanding body of literature linking air pollution with cognitive decline and dementia, several key questions remain unresolved:

1. At what point in the life course does air pollution exposure have the most significant cognitive impact?
2. Which specific pollutants or pollutant components are most detrimental?
3. Given the heterogeneity of dementia subtypes, which are most affected by air pollution exposure? Addressing these gaps is essential for advancing the field and informing public health strategies.

Furthermore, existing studies where the aim has been to investigate sensitive timing of exposure have often fallen short, due to a lack of statistical power to distinguish between different time windows, or due to a true lack of differences between the time windows investigated. While there have been promising attempts to assess air pollution exposure over the life course, and compelling evidence suggests that air pollution may influence brain development from the fetal stage—potentially even earlier through epigenetic mechanisms—the current state of knowledge in this area remains limited. There is significant room for improvement in understanding the long-term cognitive impacts of air pollution. To advance this understanding, future research requires access to studies with well-characterized lifetime exposure data, alongside detailed information on lifestyle, sociodemographic factors, and health throughout the life course. Fortunately, soon, Scandinavian countries and perhaps others too, may offer valuable insights. With historical air pollution exposure estimates dating back to the 1970s, coupled with comprehensive national registers, these resources will likely enable a more detailed exploration of how air pollution affects cognitive health over time. This will be an important step in addressing this complex and intriguing research question.

This chapter focuses specifically on long-term exposure to air pollution, rather than short-term fluctuations. Long-term exposure generally refers to cumulative pollution over extended periods, often measured in years, with various methods used across studies to estimate average exposure in relation to disease risk—in this case, dementia. While some research exists on short-term pollution variations and dementia, these studies typically examine not the onset of new cases, but the exacerbation of symptoms. For example, they might analyse emergency department visits for dementia patients, correlating the number of visits with daily changes in air pollution or temperature. These studies, though fewer in number compared to those on long-term exposure, suggest a possible link between short-term air pollution spikes and increased emergency visits. While such connections are well-established for cardiovascular and respiratory conditions, much less evidence is available for dementia. It thus seems plausible that similar patterns could emerge for dementia, however, highlighting the need for further exploration in this area.

7.7 Lack of Multiexposure Studies

The lack of multi-exposure studies was also highlighted in several reviews. Current studies do provide growing evidence that air pollution may be a significant risk factor for dementia. However, the body of research connecting other environmental exposures, such as noise, pesticides, or lack of green spaces, to dementia remains far less developed. While studies are starting to explore these factors, the evidence is still limited and inconsistent. More comprehensive research is needed to fully understand how these environmental stressors contribute to dementia risk and to clarify potential mechanisms underlying their effects on cognitive decline. According to the European Environment Agency (2022), over 100 million people in Europe are exposed to noise levels exceeding EU thresholds, with road traffic being the primary source. Noise, defined as undesirable sound, is subjective and influenced by various factors, including sound properties. Oudin emphasized the need for future studies to account for multiple environmental exposures, including noise, when investigating cognitive outcomes related to air pollution [54]. Given that road traffic produces both air pollution and noise, it's crucial to consider noise when examining the relationship between air pollution and dementia, and vice versa. For example, Chen et al. identified a link between proximity to major roads and dementia risk, suggesting that noise may contribute to this association [48]. The potential mechanisms linking noise to dementia are not fully understood. Chronic noise exposure can cause stress and annoyance, potentially leading to an overactive hypothalamic–pituitary–adrenal axis and sleep disturbances. These stress responses may increase the risk of cardiovascular diseases, which are associated with VaD and could accelerate AD pathology. Additionally, studies have shown that early exposure to traffic noise can reduce gray matter volume in adolescents. Animal studies provide further insight, demonstrating that chronic noise exposure can reduce brain volume and neuronal density and increase amyloid-beta pathology. Moreover, it has been suggested that prolonged noise exposure may lead to increased tau phosphorylation in the hippocampus.

In summary, noise may contribute to cardiovascular disease, disrupt sleep, and impair cognitive function, while greenness may promote cognitive and general health. Given the high correlation between air pollution, noise, and greenness, it is critical to study these factors together in future epidemiologic research to better understand their combined influence on AD and cognitive decline. While the plausibility of a causal link between air pollution and AD is growing, studies employing a multi-exposure approach—considering factors like noise and greenness—remain rare. These environmental factors are often correlated with air pollution, making it difficult to determine whether the observed associations with dementia are partially confounded by noise or green space exposure. Some evidence suggests noise and greenness affect cognitive function, but the literature is inconsistent, making it crucial to explore whether they independently contribute to cognitive decline. Many have recognized the accumulating evidence supporting air pollution as a risk factor for dementia. However, to establish a more robust causal link between air pollution, noise, greenness, and cognitive outcomes, future studies must adopt a multi-exposure approach. Additionally, continued efforts to identify the most harmful pollution sources will be essential in advancing our understanding. The reviews also highlight the need to better understand the specific sources and types of air pollutants that may contribute to dementia risk. There is an increasing number of studies where several measures of air pollution are included in the same models, to disentangle, e.g. effects of $PM_{2.5}$ from those of NO_2. Many studies are still single-pollutant studies, though. It seems quite clear that the most evidence is available for $PM_{2.5}$ to be associated with dementia risk compared to other measures of air pollutants, but it is still not known what components in $PM_{2.5}$ drive the effect. A key question is whether particles from different sources vary in health effects, e.g. if vehicle exhaust is more harmful than particles from domestic wood burning? There is very limited research on this, but a study by [68] did not find that particles from wood smoke were any less harmful than those from vehicle emissions in relation to dementia risk.

The need for studies in high-exposure areas, or LMICs was furthermore highlighted. Given the differences in pollution exposure levels and cumulative risks, it's critical to avoid generalizing findings from high-income countries to LMICs without further research.

In summary, there is a need for comprehensive research that considers multiple exposure sources and in high-level areas, to clarify the causal pathways involved.

7.8 Role of Dementia Subtype and Cardiovascular Disease

It is becoming increasingly important to stratify analyses of air pollution's impact on dementia by subtype to better understand the potential role of cardiovascular factors in mediating this risk. Since air pollution is strongly linked to cardiovascular diseases, it is plausible that cardiovascular dysfunction could act as a key pathway through which air pollution influences dementia, particularly VaD. Stratifying by subtype would clarify whether the risk factors differ between AD and VaD, and whether cardiovascular factors mediate these associations differently. Some studies suggest that preexisting conditions like heart failure, ischemic heart disease, and stroke are linked to stronger associations between air pollution and dementia, with stroke appearing to be the most influential intermediate condition. In one study, stroke accounted for 49.4% of air pollution-related dementia cases [66]. However, other studies have found that adjusting for cardiovascular comorbidities did not significantly alter the estimates [46]. A review noted that studies using active case ascertainment showed higher hazard ratios than those using passive case ascertainment [50], further underscoring the importance of detailed investigations into the varying roles of cardiovascular diseases for AD and VaD. By focusing on dementia subtypes, researchers could uncover specific mechanisms, such as how particulate matter and pollutants lead to vascular damage, which may be more relevant to VaD. Additionally, subtyping could help identify whether other pathways, like neuroinflammation and oxidative stress, are more dominant in AD. This approach would provide a clearer understanding of the distinct risks and pathways through which air pollution affects different types of dementia.

7.9 Conclusion

Given the profound impact of AD on individuals and society, establishing a causal association between ambient air pollution and AD would represent a significant global public health concern. The burden of AD not only affects patients and their families but also places substantial demands on healthcare systems and economic resources. According to the Alzheimer's Association, the total cost of care for individuals with AD is projected to reach over $1 trillion by 2050 in the USA alone (Alzheimer's Association, 2023). Attempts have been made to quantify the societal costs for dementia attributed to air pollution in Sweden using an estimate from a meta-analysis for a health impact assessment, but these results are heavily reliant on the assumed concentration–response function, around which there are still quite large uncertainties [67].

If future studies employing a multi-exposure approach were to confirm associations between air pollution and AD, the evidence supporting air pollution as a causal risk factor for AD would be greatly strengthened. Such findings would suggest that the health costs attributed to air pollution have been significantly underestimated, as current regulatory decisions primarily focus on mortality related to air pollution rather than its broader cognitive and neurological impacts.

Moreover, establishing a causal pathway between air pollution and AD would highlight air pollution exposure as one of the few modifiable risk factors for the disease that is prevalent in many populations. This underscores the potential for public health interventions aimed at reducing air pollution to serve as preventive measures against AD. Implementing strategies to improve air quality could not only lower the incidence of AD but also contribute to better overall cognitive health and well-being across populations. In light of the growing body of evidence linking air pollution to various health outcomes, including cognitive decline, future research must continue exploring this relationship. By focusing on the multifaceted effects of environmental exposures, we can better understand how to mitigate the risks associated with air pollution and ultimately enhance public health outcomes.

Research linking ambient air pollution and noise to cognitive decline and dementia has been gradually accumulating, but the evidence remains inconclusive, particularly regarding chronic noise exposure. A life course perspective is increasingly acknowledged as vital for identifying key determinants of cognitive decline with aging. To strengthen the evidence regarding environmental exposures, future studies should measure and model exposure to environmental factors in both early and midlife while consistently recording cognitive decline in a standardized manner. The development of cost-effective and non-invasive neuroimaging techniques or blood-based biomarkers of neurodegeneration could significantly enhance this research. Current longitudinal studies face challenges due to inadequate historical address data before participant enrolment and insufficient air pollution monitoring networks prior to 1990. Continued follow-up of existing adult cohorts over the coming decades will present opportunities for assessing late-life health outcomes, including cognitive function.

Large, population-representative studies are essential to developing a comprehensive understanding of the relationship between environmental exposures and cognitive aging across diverse populations and contexts. This research will elucidate the impact of environmental factors on vulnerable aging populations, especially in high-risk or high-pollution environments. Addressing environmental pollution will necessitate collective efforts and long-term policy initiatives at both the national and local levels. Given the scale of exposure, there is significant potential for preventive action.

References

1. Livingston, G., et al. (2024). Dementia prevention, intervention, and care: 2024 report of the Lancet Standing Commission. *The Lancet, 404*(10452), 572–628.
2. Ru, M., et al. (2021). Exploration of the global burden of dementia attributable to PM2. 5: What do we know based on current evidence? *GeoHealth, 5*(5), e2020GH000356.
3. Livingston, G., et al. (2020). Dementia prevention, intervention, and care: 2020 report of the lancet commission. *The Lancet, 396*(10248), 413–446.
4. Association, A.S. (2019). 2019 Alzheimer's disease facts and figures. *Alzheimer's & Dementia, 15*(3), 321–387.
5. Reitz, C., & Mayeux, R. (2014). Alzheimer disease: Epidemiology, diagnostic criteria, risk factors and biomarkers. *Biochemical Pharmacology, 88*(4), 640–651.
6. Gottesman, R. T., & Stern, Y. (2019). Behavioral and psychiatric symptoms of dementia and rate of decline in Alzheimer's disease. *Frontiers in Pharmacology, 10*, 1062.
7. Armstrong, R. A. (2019). Risk factors for Alzheimer's disease. *Folia Neuropathologica, 57*(2), 87–105.
8. Health, U.D.O. and H. Services. (2017). *What happens to the brain in Alzheimer's disease*. National Institute on Aging.
9. Bloom, G. S. (2014). Amyloid-β and Tau: The trigger and bullet in Alzheimer disease pathogenesis. *JAMA Neurology, 71*(4), 505–508.
10. Aisen, P. S., et al. (2017). On the path to 2025: Understanding the Alzheimer's disease continuum. *Alzheimer's Research & Therapy, 9*, 1–10.
11. Huber, C. M., et al. (2017). Cognitive decline in preclinical Alzheimer's disease: Amyloid-beta versus tauopathy. *Journal of Alzheimer's Disease, 61*(1), 265–281.
12. Haaland, K., & Swanda, R. M. (2008). *Vascular dementia. Textbook of clinical neuropsychology*. Taylor & Francis.
13. Venkat, P., Chopp, M., & Chen, J. (2015). Models and mechanisms of vascular dementia. *Experimental Neurology, 272*, 97–108.
14. Bodokia, T., et al. (2024). About vascular dementia (VAD). *Endless Light in Science, 23*(23 август extra), 12–14.
15. Desmond, D. W. (2004). The neuropsychology of vascular cognitive impairment: Is there a specific cognitive deficit? *Journal of the Neurological Sciences, 226*(1–2), 3–7.
16. Chui, H. C. (2007). Subcortical ischemic vascular dementia. *Neurologic Clinics, 25*(3), 717–740. vi.
17. Davidson, C. G., et al. (2023). Investigation into the vascular contributors to dementia and the associated treatments. *Exploration of Neuroscience, 2*(5), 224–237.
18. Custodio, N., et al. (2017). Mixed dementia: A review of the evidence. *Dementia & Neuropsychologia, 11*(4), 364–370.
19. Better, M. A. (2023). Alzheimer's disease facts and figures. *Alzheimers Dement, 19*(4), 1598–1695.
20. Iadecola, C. (2013). The pathobiology of vascular dementia. *Neuron, 80*(4), 844–866.
21. Santos, C. Y., et al. (2017). Pathophysiologic relationship between Alzheimer's disease, cerebrovascular disease, and cardiovascular risk: A review and synthesis. *Alzheimer's & Dementia: Diagnosis, Assessment & Disease Monitoring, 7*, 69–87.
22. WHO. (2021). *WHO global air quality guidelines: Particulate matter (PM2.5 and PM10), ozone, nitrogen dioxide, sulfur dioxide and carbon monoxide*. WHO.
23. Forastiere, F., et al. (2024). Choices of morbidity outcomes and concentration–response functions for health risk assessment of long-term exposure to air pollution. *Environmental Epidemiology, 8*(4), e314.
24. Chen, J., & Hoek, G. (2020). Long-term exposure to PM and all-cause and cause-specific mortality: A systematic review and meta-analysis. *Environment International, 143*, 105974.
25. Tang, J., et al. (2023). Association of air pollution with dementia: A systematic review with meta-analysis including new cohort data from China. *Environmental Research, 223*, 115048.

26. Boogaard, H., et al. (2022). Long-term exposure to traffic-related air pollution and selected health outcomes: A systematic review and meta-analysis. *Environment International, 164*, 107262.
27. Calderón-Garcidueñas, L., et al. (2002). Air pollution and brain damage. *Toxicologic Pathology, 30*(3), 373–389.
28. Calderón-Garcidueñas, L., et al. (2011). Air pollution is associated with brainstem auditory nuclei pathology and delayed brainstem auditory evoked potentials. *International Journal of Developmental Neuroscience, 29*(4), 365–375.
29. Calderón-Garcidueñas, L., et al. (2010). Urban air pollution: Influences on olfactory function and pathology in exposed children and young adults. *Experimental and Toxicologic Pathology, 62*(1), 91–102.
30. Calderón-Garcidueñas, L., et al. (2016). Interactive and additive influences of Gender, BMI and Apolipoprotein 4 on cognition in children chronically exposed to high concentrations of PM2.5 and ozone. APOE 4 females are at highest risk in Mexico City. *Environmental Research, 150*, 411–422.
31. Calderón-Garcidueñas, L., et al. (2012). Neuroinflammation, hyperphosphorylated tau, diffuse amyloid plaques, and down-regulation of the cellular prion protein in air pollution exposed children and young adults. *Journal of Alzheimer's Disease, 28*(1), 93–107.
32. Calderon-Garciduenas, L., et al. (2003). DNA damage in nasal and brain tissues of canines exposed to air pollutants is associated with evidence of chronic brain inflammation and neurodegeneration. *Toxicologic Pathology, 31*(5), 524–538.
33. Calderón-Garcidueñas, L., et al. (2008). Air pollution, cognitive deficits and brain abnormalities: A pilot study with children and dogs. *Brain and Cognition, 68*(2), 117–127.
34. Calderon-Garciduenas, L., et al. (2004). Brain inflammation and Alzheimer's-like pathology in individuals exposed to severe air pollution. *Toxicologic Pathology, 32*(6), 650–658.
35. Calderon-Garciduenas, L., et al. (2008). Systemic inflammation, endothelial dysfunction, and activation in clinically healthy children exposed to air pollutants. *Inhalation Toxicology, 20*(5), 499–506.
36. Block, M. L., & Calderón-Garcidueñas, L. (2009). Air pollution: Mechanisms of neuroinflammation and CNS disease. *Trends in Neurosciences, 32*(9), 506–516.
37. Heusinkveld, H. J., et al. (2016). Neurodegenerative and neurological disorders by small inhaled particles. *Neurotoxicology, 56*, 94–106.
38. Peters, R., et al. (2019). Air pollution and dementia: A systematic review. *Journal of Alzheimer's Disease, Preprint*, 1–19.
39. Byers, A. L., & Yaffe, K. (2011). Depression and risk of developing dementia. *Nature Reviews Neurology, 7*(6), 323–331.
40. Cao, T., et al. (2024). The relationship between air pollution and depression and anxiety disorders—A systematic evaluation and meta-analysis of a cohort-based study. *International Journal of Social Psychiatry, 70*(2), 241–270.
41. Corder, E. H., et al. (1993). Gene dose of apolipoprotein E type 4 allele and the risk of Alzheimer's disease in late onset families. *Science, 261*(5123), 921–923.
42. Juul Rasmussen, I., & Frikke-Schmidt, R. (2023). Modifiable cardiovascular risk factors and genetics for targeted prevention of dementia. *European Heart Journal, 44*(28), 2526–2543.
43. Di Battista, A. M., Heinsinger, N. M., & Rebeck, G. W. (2016). Alzheimer's disease genetic risk factor APOE-ε4 also affects normal brain function. *Current Alzheimer Research, 13*(11), 1200–1207.
44. Jung, C.-R., Lin, Y.-T., & Hwang, B.-F. (2015). Ozone, particulate matter, and newly diagnosed Alzheimer's disease: A population-based cohort study in Taiwan. *Journal of Alzheimer's Disease, 44*(2), 573–584.
45. Wu, Y. C., et al. (2015). Association between air pollutants and dementia risk in the elderly. *Alzheimer's & Dementia: Diagnosis, Assessment & Disease Monitoring, 1*(2), 220–228.
46. Oudin, A., et al. (2016). Traffic-related air pollution and dementia incidence in Northern Sweden: A longitudinal study. *Environmental Health Perspectives, 124*(3), 306.

47. Chen, H., et al. (2017). Exposure to ambient air pollution and the incidence of dementia: A population-based cohort study. *Environment International, 108*, 271–277.
48. Chen, H., et al. (2017). Living near major roads and the incidence of dementia, Parkinson's disease, and multiple sclerosis: A population-based cohort study. *The Lancet, 389*(10070), 718–726.
49. Power, M. C., et al. (2016). Exposure to air pollution as a potential contributor to cognitive function, cognitive decline, brain imaging, and dementia: A systematic review of epidemiologic research. *Neurotoxicology, 56*, 235–253.
50. Weuve, J., et al. (2021). Exposure to air pollution in relation to risk of dementia and related outcomes: An updated systematic review of the epidemiological literature. *Environmental Health Perspectives, 129*(9), 096001.
51. Killin, L. O., et al. (2016). Environmental risk factors for dementia: A systematic review. *BMC Geriatrics, 16*(1), 1–28.
52. Béjot, Y., et al. (2018). A review of epidemiological research on stroke and dementia and exposure to air pollution. *International Journal of Stroke, 13*(7), 687–695.
53. Paul, K. C., et al. (2019). Ambient air pollution, noise, and late-life cognitive decline and dementia risk. *Annual Review of Public Health, 40*, 203–220.
54. Oudin, A. (2020). Short review: Air pollution, noise and lack of greenness as risk factors for Alzheimer's disease-epidemiologic and experimental evidence. *Neurochemistry International, 134*, 104646.
55. Yu, X., et al. (2020). Exposure to air pollution and cognitive impairment risk: A meta-analysis of longitudinal cohort studies with dose-response analysis. *Journal of Global Health, 10*(1), 010417.
56. Livingston, G., et al. (2017). Dementia prevention, intervention, and care. *The Lancet, 390*(10113), 2673–2734.
57. Fu, P., & Yung, K. K. L. (2020). Air pollution and Alzheimer's disease: A systematic review and meta-analysis. *Journal of Alzheimer's Disease, 77*(2), 701–714.
58. Delgado-Saborit, J. M., et al. (2021). A critical review of the epidemiological evidence of effects of air pollution on dementia, cognitive function and cognitive decline in adult population. *Science of the Total Environment, 757*, 143734.
59. Zhao, Y.-L., et al. (2021). Environmental factors and risks of cognitive impairment and dementia: A systematic review and meta-analysis. *Ageing Research Reviews, 72*, 101504.
60. Cheng, S., et al. (2022). Long-term particulate matter 2.5 exposure and dementia: A systematic review and meta-analysis. *Public Health, 212*, 33–41.
61. Chandra, M., et al. (2022). Air pollution and cognitive impairment across the life course in humans: A systematic review with specific focus on income level of study area. *International Journal of Environmental Research and Public Health, 19*(3), 1405.
62. Abolhasani, E., et al. (2023). Air pollution and incidence of dementia: A systematic review and meta-analysis. *Neurology, 100*(2), e242–e254.
63. Wilker, E. H., Osman, M., & Weisskopf, M. G. (2023). Ambient air pollution and clinical dementia: Systematic review and meta-analysis. *BMJ, 381*, e071620.
64. Segersson, D., Johansson, C., & Forsberg, B. (2021). Near-source risk functions for particulate matter are critical when assessing the health benefits of local abatement strategies. *International Journal of Environmental Research and Public Health, 18*(13), 6847.
65. Russ, T. C., et al. (2021). Life course air pollution exposure and cognitive decline: Modelled historical air pollution data and the Lothian Birth Cohort 1936. *Journal of Alzheimer's Disease, 79*(3), 1063–1074.
66. Grande, G., et al. (2020). Association between cardiovascular disease and long-term exposure to air pollution with the risk of dementia. *JAMA Neurology, 77*(7), 801–809.
67. Kriit, H. K., et al. (2021). Annual dementia incidence and monetary burden attributable to fine particulate matter (PM2. 5) exposure in Sweden. *Environmental Health, 20*(1), 1–12.
68. Oudin A, Segersson D, Adolfsson R, Forsberg B. Association between air pollution from residential wood burning and dementia incidence in a longitudinal study in Northern Sweden. PLoS One. 2018 Jun 13;13(6):e0198283. https://doi.org/10.1371/journal.pone.0198283. PMID: 29897947; PMCID: PMC5999109.

Chapter 8
Air Pollution and Mental Health

Tara J. Crandon, Siti Nurshahida Nazli, and Dwan Vilcins

Keywords Air pollution · Climate change · Mental health · Wellbeing · Public health

8.1 Introduction

8.1.1 Air Pollution Risks to Human Health

The air we breathe is essential for life, yet it also poses a major threat to human health. Air pollution, the presence of harmful substances in the atmosphere, is a global crisis. From smog-choked cities to the invisible particulate matter lingering in the air of even the most pristine environments, poor air quality carries far-reaching consequences across social, economic, and health domains. Air pollution is a global risk factor for mortality and was the second leading cause in 2021, following high blood pressure in adults and malnutrition in children [36].

Air pollution arises from a wide range of natural and man-made sources. Major contributors include household combustion devices (i.e., wood-burning stoves and fireplaces), motor vehicle emissions, industrial facilities, and smoke from wildfires [47]. Such sources produce a complex mixture of pollutants [99]. Among the most

T. J. Crandon (✉)
The School of Public Health, The University of Queensland, Herston, QLD, Australia
e-mail: t.crandon@uq.edu.au

S. N. Nazli
The University of Queensland, Child Health Research Centre, South Brisbane, QLD, Australia

Faculty of Health Sciences, Universiti Teknologi MARA Cawangan Pulau Pinang, Pulau Pinang, Malaysia

D. Vilcins
The University of Queensland, Child Health Research Centre, South Brisbane, QLD, Australia

K. M. Kanninen, A. R. White (eds.), *Air Pollution and the Brain*,
https://doi.org/10.1007/978-3-031-99302-2_8

concerning are particulate matter (PM)—microscopic solid or liquid particles suspended in the air—and carbon monoxide, an odorless, colourless gas resulting from incomplete combustion. This pollutant is generated by household energy use, construction and manufacturing, unpaved roads, fields, industry, the agriculture and transport sectors, coal-fired power plants, sand and desert dust and waste burning, and deforestation [33, 34, 98]. Gaseous pollutants also pose significant concerns, such as ozone, a reactive gas formed through chemical reactions involving other pollutants, nitrogen dioxide, a reddish-brown gas with a pungent odour, and sulphur dioxide, a colourless gas with a sharp, irritating smell.

A plethora of research shows that air pollutants can have a profound biological impact, leading to the exacerbation or development of poor physical health and contributing to millions of premature deaths annually [5, 65]. Certain subpopulations are disproportionately affected, including children and young people, pregnant women, the elderly, and residents of low-income communities [99]. The risk of exposure to poor air quality is higher for populations in low- and middle-income countries, who are exposed to 1.3–4 times higher levels of ambient PM2.5 on average [36]. Emerging in the literature is an understanding of how air pollutants may impact mental health, and this chapter will provide an overview of this work. Exploring the complexities of air pollution and its extensive repercussions allows us to grasp the urgency of tackling this issue, whilst also outlining potential areas to intervene.

8.2 Air Pollution and Mental Health

Beyond physical health implications, the impacts that air pollution can have on mental health and well-being are becoming increasingly apparent. Mental health can be thought of as one's emotional and psychological state, encompassing their ability to cope with daily life as well as adversity [73]. It is important to note that mental health exists on a spectrum, fluctuating throughout life, and shaped by the environments within which people live. When mental health deteriorates to a significant degree, there may be significant negative impacts on personal, occupational, and social functioning. For some, poor mental health may meet the criteria for a mental health disorder. For example, Major Depressive Disorder or Generalised Anxiety Disorder, as outlined in the Diagnostic and Statistical Manual for Mental Health Disorders (DSM-V-TR) [2].

As mentioned, environmental influences can shape and influence mental health. Air pollution, in particular, is being increasingly identified as an adverse environmental factor. Globally, daily and long-term exposure to air pollution has been associated with increased odds of mental health disorder incidence, such as depressive and anxiety disorders, higher rates of psychological distress, suicidal ideation, and substance use problems, as well as increased admissions to psychiatric emergency units [15, 35, 67, 73, 85]. While evidence in this space is accumulating, establishing causality between air pollution and mental health is a difficult task [114]. Indeed, air pollution may impact mental health in complex direct and indirect ways and must therefore be examined alongside potential influential variables. The following sections will look to summarise current evidence on air pollution and mental health, noting individual impacts as well as broader community risk factors (see Fig. 8.1 for an overview).

8.3 Mental Health Effects

8.3.1 Depressive Disorders

Much of the evidence on the adverse effects of air pollution on mental health focuses on symptoms of depression, such as low mood, irritability, sleep disturbances, diminished energy, and loss of interest and pleasure [9]. A meta-analysis demonstrated that short-term and long-term exposure to a variety of air pollutants increased the risk of depression [115]. In the short term, depressive symptoms may be higher on days with extreme air pollution levels [14]. However, exposure to air pollution may also have prolonged impacts on mental health, with one study showing that the incidence of depression may increase 10 years following a higher average level of exposure to air pollution in certain populations (e.g., Indonesians) [114]. Furthermore, a study on adolescents found estimates of air pollution exposure at age 12 were associated with increased odds of major depressive disorder at age 18 [82].

The incidence of depression and antidepressant use has been specifically linked to increased concentrations of particulate matter, nitrogen dioxide, and sulphur dioxide [95, 116, 117] . Of importance, it has been shown that even short-term exposure to nitrogen dioxide was associated with depression [118]. Some authors suggest higher levels of air pollution exposure may be associated with symptom severity, with an increased likelihood of clinically significant depression in women specifically [114]. Populations identified as being particularly predisposed to increased risk of depression from air pollution exposure include men, pregnant women, elderly people, people with pre-existing health problems, people using medication such as antidepressants, and people from lower-income backgrounds [76, 116, 117, 119, 120, 122].

8.3.2 Psychosis and Risk

Psychotic experiences reflect a disconnection with reality, wherein people may experience hallucinations, delusions, and disorganised speech or thought patterns. There are several types of psychotic disorders in the DSM-V-TR (e.g., schizophrenia, schizoaffective disorder), yet psychotic experiences can also be present for those with a mood disorder such as depression or bipolar disorder [2]. There is growing evidence that air pollution may be a risk factor for schizophrenia and

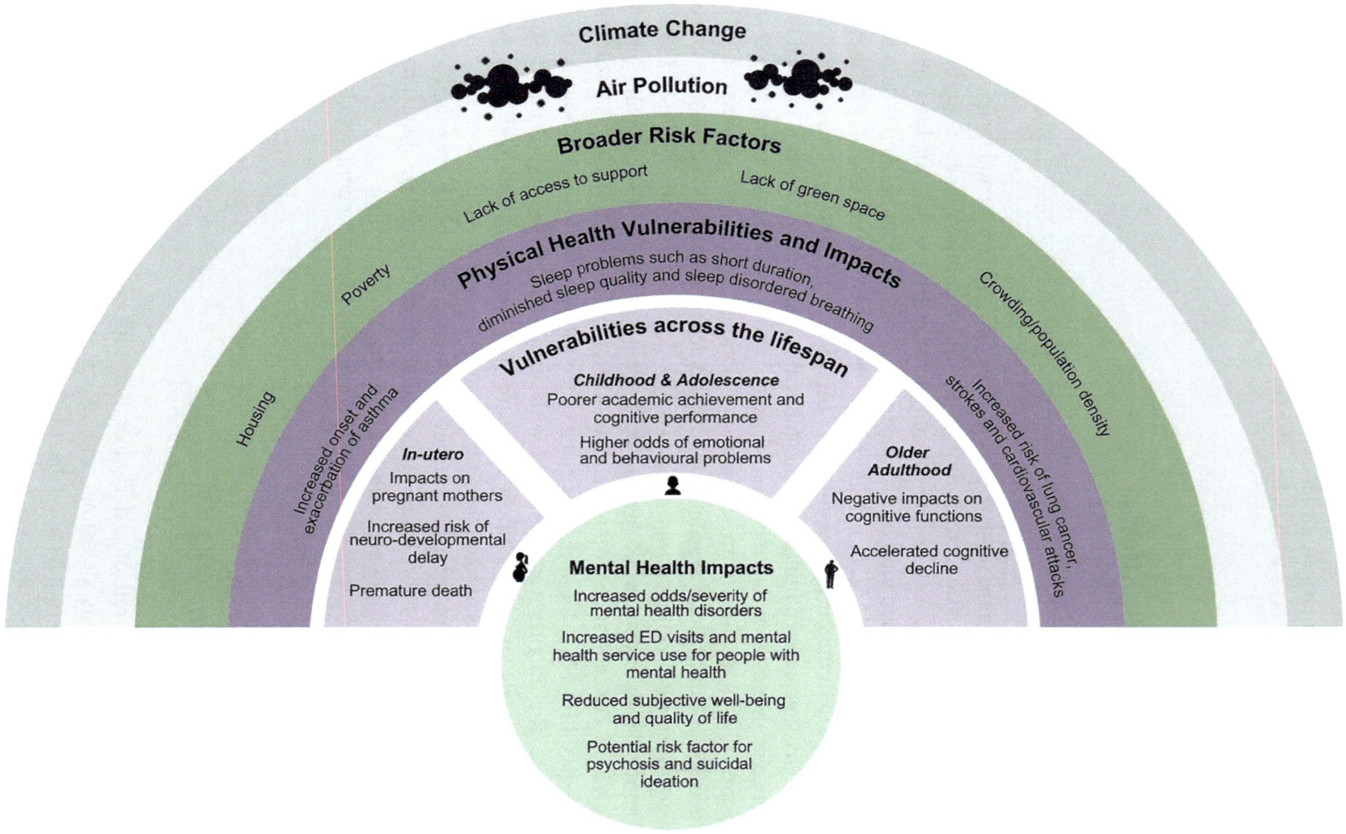

Fig. 8.1 An overview of the impacts of air pollution on mental health

psychotic experiences [3, 90, 110], with one study showing gaseous pollutants may present an immediate risk and particulates may present a delayed risk [90]. The risk of suicide and suicidal ideation may also be influenced by air pollution, particularly for short-term increases in the level of particulate matter [123]. A study in Korean adults also found associations between long-term exposure to ambient air pollution and stress, depression, and suicidal ideation, particularly for men and elderly adults [116].

8.3.3 Emergency Department Visits and Mental Health Service Use

The emergency department is a critical source of care for people experiencing mental health problems, particularly in times of crisis when the ability to care for oneself or function effectively is compromised, or when symptoms place a person at risk to themselves or others. Substantial evidence links exposure to air pollution to increases in psychiatric emergency room visits, wherein people may present due to depression or anxiety, psychosis, neurosis, stress, substance use, bipolar disorder, schizophrenia, self-harm, and suicidal risk [29, 38, 62, 68, 74, 107, 108]. Mental health service use also increases amongst those exposed to higher levels of air pollutants who have also been diagnosed with a psychotic or mood disorder [67]. Much of this evidence is primarily on particulate matter, suggesting that increased concentrations of particulate matter are linked to acute worsening of mental health problems [29, 62, 108].

8.3.4 Poor Sleep

Sleep is a biological necessity, serving as an essential regulatory function for brain performance, physical health, and mental health [70]. Healthy sleep depends on a number of factors such as adequate duration, quality, timing, and regularity. Problems in these areas are linked to poor mental health, whereby poor sleep can contribute to mental health problems or vice versa [70, 79, 88]. Evidence suggests greater exposure to both indoor and ambient air pollution is associated with poor sleep health, with impacts including shorter duration of sleep, diminished sleep quality, as well as sleep disordered breathing (i.e., sleep apnea) [18, 53]. Some authors have suggested this relationship can be explained by central nervous system, allergic, and non-allergic mechanisms [124]. A systematic review also highlighted that the relationship between obstructive sleep apnea and air pollution varied according to the type of air pollutant, as well as across seasons, temperature, and countries [18]. Irrespective of this, poorer sleep health resulting from air pollutant exposure is likely to interact with and/or exacerbate negative outcomes for physical and mental health [121].

8.3.5 *Stress and Anxiety Disorders*

Though not inherently dysfunctional, anxiety can become a clinical problem (e.g., social anxiety disorder, generalised anxiety disorder) when it is significant, persistent, and uncontrollable. Greater exposure to air pollution has been linked to increased prevalence of anxiety symptoms and disorders [113], though this relationship has not been consistent across all studies [95]. Long-term exposure to particulate matter has been linked to anxiety symptoms and stress [123]. In elderly women, the odds of experiencing anxiety symptoms were found to be greater when exposed to higher levels of particulate matter across multiple periods of time (i.e., 1 month, 12 months) [125]. A study conducted in three subtropical cities in China found that short-term exposure to all air pollutants was associated with anxiety disorders [117].

8.4 Vulnerability Across the Lifespan

As mentioned, the adverse impacts of air pollution on health (whether physical, social, or psychological) are not evenly distributed amongst populations [55]. Populations that are biologically susceptible to air pollution may similarly be at greater risk of negative mental health effects, such as pregnant women and fetuses, children and adolescents, elderly people, those with physical health problems, as well as people living in low socio-economic conditions [55]. One way to examine how these groups are vulnerable is to take a lifespan approach.

8.4.1 *Pregnancy/In Utero and Infancy*

During the perinatal period (i.e., from conception up until approximately a year after birth), women are particularly vulnerable to the development or exacerbation of mental health problems [21, 91]. Biological (e.g., epigenetic modifications, low vitamin D plasma levels) and psychosocial changes (e.g., interpersonal conflict, isolation, unemployment) can particularly contribute to poor mental health in mothers, which may have flow-on effects for the mother-child bond, as well as the child's well-being and health [91]. Exposure to air pollution during this time may contribute to these risks, with studies showing associations between specific air pollutants (i.e., PM2.5 and NO_2) and the onset of Postpartum Depression (PPD) in the months after childbirth [91]. One study that tracked pregnant women at preconception, first trimester, and whole pregnancy showed that ambient air pollutants were associated with increased odds of depression and an unspecified mental disorder complicating pregnancy [109].

During pregnancy, the fetus is vulnerable to environmental toxicants like air pollutants, with potential mechanisms including compromised placental growth and function, systemic and neuroinflammation, as well as endogenous oxidative stress [1]. Air pollution may also negatively affect neurodevelopment and neurocognitive performance across the lifespan. Exposure in utero has been associated with increased risk of neuro-developmental delay, and exposure throughout early childhood has been inversely linked to neurodevelopment [19]. Furthermore, air pollution has been associated with post-neonatal mortality (between 28 days and the first year of life), particularly due to respiratory causes and sudden infant death syndrome [32]. Losing a child has a profound impact on the well-being and health of caregivers and families, potentially leading to increased hospital admissions for psychiatric or medical concerns, relationship breakdown (e.g., divorce), greater physical and mental health problems, and higher mortality [96].

8.4.2 Childhood and Adolescence

Children and adolescents exposed to air pollution over time may be at increased risk of anxiety and developing cognitive or behavioural challenges [10, 61]. More specifically, a review across seventeen studies found that young people (i.e., aged 10–24 years) exposed to air pollution (e.g., second-hand smoke, traffic emissions, ozone) were more likely to have elevated symptoms of depression, generalised anxiety, psychosis, impulse control and conduct disorder problems [92]. These effects may be greatest for young people in the highest quartile of pollutant exposure. In an Australian longitudinal study, exposure to moderate and high particulate matter in pregnancy, first year of life, and over a child's life was associated with higher odds of emotional and behavioural problems (e.g., hyperactivity/inattention, emotional symptoms, peer problems), as well as gross motor delay, compared to children with low levels of exposure [1]. Exposure to elemental carbon attributable to traffic at birth was associated with increased anxiety and depressive symptoms reported by children aged 12 years [106]. Furthermore, children below the age of 15 living in a rural region of Spain were predisposed to greater mental health problems when exposed to air pollution, namely behavioural and developmental disorders (nitrogen dioxide, ozone and sulphur dioxide), anxiety (particulate matter, ozone and sulphur dioxide), as well as ADHD and eating disorders (general air quality) [61]. In adolescents aged 18 years, evidence suggests psychotic experiences are more common in those with the highest levels of annual exposure to nitrogen dioxide [66]. As it relates to cognitive functioning, negative outcomes linked to air pollution include poorer academic achievement and cognitive performance in memory and attention tasks [19, 64]. Some studies do observe contradictory findings to the above, observing no relation between exposure to air pollution and depressive or anxiety symptoms in children [16, 43]. These inconsistent findings could in part be explained by diverse methods of capturing air pollution estimates, varying time periods in which the study was conducted (e.g., over 1 year compared to over several years), as well as using different psychometric instruments to assess anxiety and depressive symptoms in children [16, 43].

8.4.3 Older Adulthood

In older adults, exposure to air pollution has been associated with anxiety and depressive symptoms [52, 76, 112]. One study found that long-term exposure to particulate matter, nitrogen dioxide, and ozone was associated with an increased risk of being diagnosed with late-onset depression [78]. Poor mental health may be especially prevalent in older adults who have physical health problems (e.g., poor fitness and history of problems like respiratory issues or stroke) or those with a low socio-economic background [76, 112]. Exposure to air pollution in later life has also been associated with negative impacts on cognitive functions (e.g., immediate recall, mental intactness) as well as accelerated cognitive decline [19, 48], with some studies showing an adverse relationship between air pollutants and dementia [7, 126].

8.5 Possible Mechanisms

8.5.1 Biological and Physical Health

Multiple biological explanations have been given regarding the links between air pollution and poor mental health, including neurological changes, oxidative and nitrosative stress, systemic and neuroinflammation, and changes in hormone production or hormone dysregulation [8]. Furthermore, being exposed to air pollution long-term can have deleterious effects on physical health problems, such as increasing the onset and exacerbation of asthma, increasing the risk of lung cancer, strokes, and cardiovascular attacks [94]. Chronic exposure to air pollution has also been associated with pulmonary and metabolic dysfunction, contributing to higher rates of premature mortality [127]. These types of health problems and conditions are known to have flow-on effects for mental health and well-being. People with asthma, e.g., are more predisposed to have a mood disorder such as anxiety, depression, or bipolar [128, 129]. Yet also worth acknowledging are the various ways in which air pollution and associated physical health problems may burden mental health, either through financial pressures or added strains to social, occupational, and personal life domains.

A plethora of research shows that mental and physical health are bidirectionally linked, and this relationship may be moderated by air pollution. For example, in Korean patients with asthma, cardiovascular disease, or diabetes mellitus, air pollution was found to increase the rate of emergency department visits for depressive episodes [130]. Several authors have further suggested that physical health may be a mediator for the links between air pollution and mental health. For example, one study showed that physical health mediated the effects of air pollution on depression. However, physical health was not a mediator between air pollution and subjective levels of happiness [131].

8.5.2 Subjective Well-Being and Quality of Life

Though not all air particulates can be seen with the naked eye, visibility reduction (i.e., haze or smog) is perhaps the most apparent indicator of air quality. Haze subdues colour and contrast, which authors suggest can negatively impact the aesthetics and enjoyability of the environment [39, 41]. The environment can shape subjective happiness and life satisfaction, which are important indicators of short and long-term well-being. Because of this, it is perhaps unsurprising that an awareness of the presence of air pollution is being increasingly implicated as a factor that may affect people's mood and negatively impact subjective well-being [51]. In China, e.g., air pollution has been found to negatively impact subjective levels of happiness directly, and this relationship was also found to be mediated by health [51]. Furthermore, residents living in the city of Tangshan (an urban area with particularly poor air quality in China) describe anxiety, helplessness, and uncertainty regarding smog, with mothers particularly reporting concern over their children's health [50]. Members of a rural community in the United States, who were particularly impacted by prolonged wildfire smoke, similarly reported heightened anxiety and depressive symptoms, isolation, and concern for children [40]. In Germany, another study found that less air pollution was associated with greater self-esteem in women, as well as higher satisfaction in life and higher stress resilience in a general adult population [73].

8.5.3 Risk Perceptions

Perceptions of risk play an important role in shaping how people respond to stressors or adversity, thereby impacting mental health and well-being [69]. When risk is determined to be high, this can evoke feelings like anxiety, stress, and sadness, but also potentially lead to protective behaviours. Conversely, stressors deemed to be low risk may not yield a protective behavioural or coping response. As mentioned, there are multiple sources of air pollution, including industrial emissions, household heating and cooking, wildfires as well as dust. Perception and knowledge over the impacts of these different sources on health can vary, wherein underestimation of these risks may mean people do not take appropriate measures to mitigate or protect themselves from air pollution [69]. Indeed, several studies have found that some people are cognisant of the adverse impacts of air pollution on health, yet also diminish or even deny the impacts on their own health [69, 80]. For others, the ability or capacity to address air pollution may be limited by other factors known to affect well-being and quality of life (e.g., poverty). For example, residents living in two slum areas of Nairobi reported feelings of helplessness and frustration regarding the ability to address air pollution prevalent within the community, due to practical (e.g., non-ventilated homes), social (e.g., conflict with neighbours), and systemic barriers [63]. For those who are unable to decrease or reduce their exposure to air pollution, this is where emotional regulation strategies may become particularly important. One study, e.g., showed that when people who tend to suppress their emotions used cognitive reappraisal (i.e., changing how one thinks about or appraises a situation), they experienced less overall smog-induced psychopathological symptoms [14].

On the other hand, one does not need direct experience or to physically perceive a threat to be impacted by it. The nocebo effect (e.g., in contrast with the placebo effect) describes the negative impacts of expectations [132, 133]. Simply being concerned about worsening air quality and pollution can evoke worry and stress, which will, in turn, impact mental health. In relation to this is the experience of 'climate anxiety', which describes fear and worry relating to the threats of climate change. As climate change unfolds, concerns about the negative impacts of air pollution comprise one aspect of climate-related anxiety and distress. Those who are especially predisposed to the negative effects of air pollution will likely be most anxious or distressed about it. For example, children or parents/caregivers of children, women, elderly people, people with physical health problems, or people living in impoverished conditions where air quality is poor and the ability to protect oneself is limited.

Though these biological and psychological understandings can help to characterise the relationship between air pollution and mental health, it is clear that the relationship is complex and multifaceted. The impact of broader community and systemic factors must also be considered, particularly when identifying areas to intervene.

8.6 Community Effects and Broader Risk Factors

8.6.1 Poverty, Poor Housing Conditions, and Lack of Access to Support

Communities that are poorer, who have limited access to clean environments, reduced access to quality healthcare, and those living in poor housing conditions, are typically exposed to higher levels of air pollution compared to wealthier populations [22, 42]. Given the links between air pollution and other socio-economic problems (e.g., poverty, housing), air pollution can be thought of as a social and structural determinant of health [71]. This is exemplified in 28,081 clusters of villages in rural areas and census enumeration blocks in urban areas in India, wherein poverty has restricted access to household amenities (e.g., electricity, improved sanitation, solid fuels for energy needs, access to safe drinking water), contributing to PM2.5 emerging as a major risk factor for the population [23]. Similarly, during 2022, about 46.3% of the rural population in Central America and 23.3% in South America had limited access to clean fuels [34]. This led to a rise of 3.9% in premature deaths due to PM2.5 from fossil fuels between 2005 and 2020, corresponding to 123.5 premature deaths per million people in Latin America.

Cao et al. [11] emphasise that people in economically disadvantaged cities face greater mental health challenges due to polluted air, and this is especially problematic when people are unable to or have limited resources to alleviate their level of exposure. Exacerbating this problem, the rise in mental health problems places additional pressure on mental health services, which are often already scarce in low-income areas [5, 11]. This can have flow-on effects for community well-being. When lacking the resources to move, mitigate their exposure levels, or effectively advocate for change, these communities may find themselves trapped in a cycle of poor health and economic hardship. As a result, environmental justice movements have become essential in highlighting the inequalities these populations endure and advocating for policy changes to address these disparities [57].

8.6.2 Household Overcrowding

Urbanicity is a risk factor for the incidence of several mental health disorders (e.g., psychotic disorders, autism spectrum disorder), and it has been suggested that this relationship could partly be explained by air pollution [74]. Kajjabo et al. [44] emphasise that overcrowding, driven by rapid urbanisation, further worsens indoor air quality. Ineffective building designs, which fail to provide adequate ventilation, exacerbate the issue because windows and doors are often kept closed, trapping pollutants inside [44]. Additionally, overcrowding can lead to an increase in moisture levels that encourage the presence of mold [37, 54]. This degradation in air quality has been associated with a range of health problems, including respiratory issues, fatigue, and allergies.

Children are particularly vulnerable to the negative effects of overcrowding. According to Solari and Mare [89], overcrowded living environments can lead to emotional and behavioural problems in children, disrupt sleep, and hinder academic performance due to limited space for studying and high levels of noise. These conditions not only affect mental and emotional well-being but also increase the risk of respiratory illnesses such as asthma and infections like acute respiratory infections (ARI) [54, 89]. Firdaus and Ahmad [28] found that overcrowded households are 62% more likely to experience ARI, with other factors like poor ventilation and indoor air pollution further exacerbating the prevalence of asthma and tuberculosis.

Broadly, overcrowding and high population density are pressing issues, with studies showing both conditions are associated with poorer mental health [27, 84, 86, 111]. Furthermore, poor living conditions as a result of overcrowding might worsen mold growth, pest infestation, poor ventilation, and exposure to chemicals like pesticides and biocides, wherein exposure to these physical health risks may also confer risks to mental health [5, 28, 54]. Such living conditions may also be found in densely populated urban areas, which are also likely to be exposed to higher levels of air pollution, potentially compounding adverse effects on mental health [5]. These findings highlight the urgent need to address overcrowding and improve housing conditions to mitigate both physical and mental health risks.

8.6.3 Lack of Green Space

Green spaces (e.g., parks, urban forests) often serve as important community hubs where people can relax and alleviate stress, socialise, exercise, and engage in recreational activities. Given these opportunities as well as the reduced exposure to air pollutants in green spaces, it is no surprise that substantial research suggests spending time in green spaces is beneficial for physical and mental health [5, 6, 30]. Communities with limited access to green spaces are especially vulnerable to mental health problems, and this issue may be further exacerbated when considering prolonged exposure to air pollution [4–6, 11]. Indeed, while higher levels of PM2.5 are associated with an increased prevalence of depression, there is a reduction of mental health problems associated with air pollution for those who live near green spaces [11, 20, 98]. One study, in particular, examined mood symptoms in elderly people after taking a walking route on different days through a 'grey' urban area (e.g., a busy street with a high degree of air pollution) and a 'green' walking route (i.e., a residential area with trees and gardens). Depressive symptoms were found to decrease significantly following the green walking route, as opposed to no symptom changes following the high air pollution walk [83]. Given that green spaces may appear to mitigate the mental health effects of air pollution, and may even reduce the air pollutants, lack of access to these spaces may translate to increased negative effects and represent a health equity issue.

8.6.4 The Overarching Impact of Climate Change

For decades, air pollution has imposed a substantial toll on both human health and the economy, contributing to premature deaths, widespread illness, decreased productivity, and weakened communities. Although policies and technological advancements have improved air quality in certain areas, proving that pollution is not an inevitable outcome of economic development, the risks of air pollution remain and are growing increasingly complex. These challenges are now further complicated not just by the aging global population, but also by climate change [36].

Research indicates that 3.6 billion people reside in regions highly vulnerable to the impacts of climate change [103]. Projections suggest that between the years 2030 and 2050, climate change will be responsible for an alarming increase in mortality rates, with an estimated 250,000 additional deaths annually. This grim toll is attributed to a range of factors exacerbated by climate change, including malnutrition, the spread of malaria, diarrheal diseases, and the severe consequences of heat stress [103]. The Lancet Countdown Latin America 2023 presented 34 indicators showing the relationship between climate change and health up to the year 2022. There are severe direct impacts due to a warming trend across all countries in Latin America, including exposure to ambient temperatures 0.38 °C higher than in 1986–2005 and a heatwave that exposed 248% of infants and 271% of elderly people from 2013 to 2022 [34]. The changes in the ecosystem further exposed people to the risk of wildfires from high fire danger and also air pollution [33, 34, 98].

There is a complex interaction between air pollution and climate change that causes threats to human health and the environment worldwide [65]. Many air pollutants are greenhouse gases that trap heat in the atmosphere, including carbon dioxide (CO_2), methane, and nitrous oxide, emitted from the burning of fossil fuels for energy, transportation, and industrial processes, are major drivers of both air pollution and climate change [75]. The significant risks posed by air pollution and climate change are further compounded by their interaction, each intensifying the other's effects. Hotter and drier conditions can elevate general air pollution levels, particularly increasing ozone through heightened photochemical production. Pollutants such as black carbon, methane, tropospheric ozone, and aerosols alter the amount of sunlight reaching Earth, causing temperatures to rise and leading to the melting of icebergs and glaciers [56]. Additionally, future, more severe heat waves driven by climate change are expected to spark more wildfires, releasing greater amounts of harmful greenhouse gases and particulate matter [75].

8.7 Interventions

Air pollution poses a significant risk to physical and mental health, and the need to reduce exposure to harmful pollutants is therefore clear. The complexity of the problem, coupled with the influence of broader risk factors, calls for a systemic approach. Considering this, the following section will outline ways in which to address the adverse effects of air pollution on mental health. Specifically, recommendations will be made for policy and prevention, public health, and community intervention, as well as potential ways the mental health field can respond and act.

8.7.1 Policy, Prevention, and Public Health

8.7.1.1 Reducing Exposure to Air Pollution

The way communities are organised, cities are planned, and societies are supported has significant implications for mental health, as these population-level factors directly influence biological, social, environmental, and psychological outcomes. Initiatives aimed at reducing air pollution, such as green transportation policies and stricter industrial regulations, have been shown to improve both physical and mental health outcomes at the community level. In cities like Brussels, where transportation-related pollution is a significant issue, these efforts could help alleviate the burdens associated with both physical illnesses and mental health [7]. There are existing cases of policies relating to air pollution yielding significant public health benefits, particularly in addressing mental health. In China, for example, the implementation of stringent air quality controls resulted in a marked decrease in pollutants such as PM2.5. This reduction was linked to lower depressive symptoms and overall improvements in mental health outcomes [105]. This highlights the dual necessity of individual precautions and robust policy measures to mitigate both the physical and mental health impacts of air pollution.

Equally important are policies that will address social issues interwoven with air pollution, such as poverty, poor housing conditions, as well as overcrowding and high population density [86]. Funding and regulating the service and maintenance of ventilation systems in public spaces (e.g., schools, shopping centres), e.g., could help to improve indoor air quality and prevent overheating [46]. Research has shown that adequate ventilation systems reduce exposure to indoor air parameters such as PM2.5, carbon monoxide, and TVOCs, and improve temperature and relative humidity levels, which are known to cause respiratory and mental health issues, such as stress, depression, and anxiety [24]. One study found that compared to individuals in poorly ventilated homes, those in homes with moderate ventilation had a 28–53% lower risk of experiencing depression, while those in homes with high ventilation saw a 43–67% reduction, regardless of demographic or lifestyle factors [24]. Moreover, improving ventilation, heating, and air conditioning in overcrowded homes can reduce moisture and mold growth [54]. Though installing chimneys or hoods can reduce smoke exposure from cooking and indoor activities, governments should further prioritise increasing access to clean technologies and helping families transition to more effective ventilation methods (e.g., cleaner fuels, low-emission stoves) [101]. Continued focus in these areas can lead to better health outcomes physically, psychologically, and socio-economically.

8.7.1.2 Messaging, Safety, Tangible Actions

A comprehensive public health strategy is necessary when communicating the risks of air pollution, as well as educating the public on ways to reduce exposure. This can be achieved through effective communication, preventive measures, and practical interventions. Messaging around air quality needs to be clear, accessible, and practical to encourage behaviour change and reduce exposure to harmful pollutants. For instance, raising awareness of air pollution risks through healthcare professionals can increase public awareness and promote preventive measures, such as limiting outdoor activities during high-pollution days [87]. Communicating safety strategies and tangible actions that can be undertaken may be particularly important for vulnerable populations such as children, the elderly, and individuals with pre-existing physical or mental health problems. Besides that, health advisories can also be shared using multiple media platforms to inform the public about air quality hazards and encourage safety practices. On an individual level, personal protective strategies can play a vital role in mitigating health risks. For example, using air quality monitoring systems, installing portable air cleaners equipped with HEPA filters, and avoiding high-exposure activities are all proven methods to reduce the dangers of air pollution [13, 87]. The rationale and application of these solutions need to be effectively conveyed, taking into account the needs, values, and resources available to people.

8.7.1.3 Training for Health Professionals

According to the WHO, it is internationally acknowledged that health professionals play a critical role in the fight for clean air. Health professionals are trusted sources of information, communicating the risks of air pollution with governmental authorities as well as the public. At the same time, health professionals will be instrumental in advocating for and supporting interventions that will reduce the physical and psychological impacts of air pollution. Yet, many health professionals are not adequately trained in the risks and consequences of air pollution, nor the strategies necessary to mitigate its effects. Health organisations and educational institutions must ensure that health professionals have clear guidelines and training to be able to communicate and respond to the risks posed by air pollution. For example, the WHO has developed the Air Pollution and Health Training (APHT) toolkit for health workers across clinical and public health sectors [100]. This toolkit equips health professionals with the knowledge to understand the health risks posed by air pollution and to identify effective risk reduction strategies. Health management plans are also critical to supporting physical as well as mental health and well-being. These plans could be implemented broadly for the community (e.g., community-based strategies for people living in urban neighbourhoods on high pollution days) or be tailored to the specific needs of individuals (e.g., extra preventer inhalers for individuals with asthma during bushfire smoke days). Additionally, collaboration between environmental scientists, social and healthcare scientists, as well as health professionals, is crucial. Given the systemic and complex ways that air pollution may impact health, these collaborations are essential to understanding underlying mechanisms, as well as impactful treatment options and primary prevention strategies [81].

8.7.1.4 Climate Change Mitigation

As climate change and fossil fuels are interconnected with air pollution, efforts to effectively and adequately address the climate crisis are critical. As mentioned, fossil fuel combustion releases harmful pollutants such as PM2.5, NO_2, and CO_2. Reducing and eliminating fossil fuel consumption will thereby reduce the presence of these pollutants, simultaneously promoting air quality and lowering greenhouse gas emissions (which significantly contribute to climate change). Research has indicated that phasing out fossil fuels could prevent premature deaths by 0.79 million (95% confidence interval: 0.75–1.8 million) from air pollution, particularly in high-pollution areas such as Asia. Thus, robust climate policies in these regions may protect the lives of millions by 2050, specifically reductions in air pollution-related mortality [104].

Further benefits of reducing fossil fuel consumption and improving air quality include decreased respiratory illnesses, decreased hospital admissions, and reduced healthcare costs, as evidenced in cities that have adopted cleaner and renewable energy sources [60, 93]. These transitions promote energy efficiency, which reduces the emission of toxic pollutants, especially in urban areas, and lessens the health burden associated with fossil fuels.

The health benefits further emphasise the importance of public health strategies that prioritise the reduction of fossil fuel emissions. This strategy not only reduces the pressure on healthcare systems, but also fosters a healthier environment, and provides long-term sustainability and protection for future generations. Additionally, adopting policies targeting CO_2 emissions brings broader social and economic advantages, particularly for vulnerable populations like children, who are disproportionately affected by both air pollution and climate change [72]. Through cooperative actions in climate policy and public health, both immediate and long-term challenges can be addressed.

8.7.2 Community and Individual Strategies

8.7.2.1 Increasing Access to Green Spaces

The importance of incorporating accessible green spaces in urban areas cannot be overstated, especially for those in densely populated or economically disadvantaged areas, as it greatly supports mental health and overall well-being. Urban green spaces include smaller green space features like street trees and roadside vegetation, green spaces not available for public access or recreational use like green roofs and facades, or green space on private grounds, as well as larger green spaces for various social and recreational functions like parks, playgrounds or greenways [102]. Not only do green spaces reduce air pollution, but they are also known to reduce physical health problems (e.g., mortality due to chronic diseases, obesity) and directly promote mental health and well-being [6, 26]. The World Health Organization (WHO) [102] has provided examples of ways that green spaces can be modified, changed, or improved in urban settings, to maximise opportunities for populations, including those from lower socio-economic backgrounds [102]. These initiatives may be environmentally, socially, physically, and psychologically beneficial for residents, though they must be integrated into broader urban strategies like master plans and sustainability policies. The key to reducing risks and maximising benefits from effective green space is effective design, implementation, and maintenance. Assessing the availability, accessibility, and quality of green spaces, as well as monitoring their usage by specific population groups (i.e., children, the elderly, working adults, pregnant women, cultural and ethnic minorities, and individuals with mental or physical health problems), is also essential for guiding evidence-based interventions [6].

8.7.2.2 Physical Activity

The environment can influence people's engagement in physical activity (e.g., exercising, playing sports). Where a good physical space may facilitate physical activity, a poor physical environment (e.g., the presence of air pollution, an overcrowded space) can hinder this behaviour [31]. Though physical activity is known to be beneficial for mood and mental health, exercising outside could also lead to an increase in the volume of inhaled air, thereby potentially augmenting exposure to air pollutants [12, 58, 77]. This is not to say that physical activity should be decreased. In fact, lack of or reduced physical activity could further impair health or exacerbate disease, and exercising indoors may not necessarily reduce exposure [25]. Authors suggest a need for health practitioners to encourage regular exercise, while adjusting the timing and location of this activity to minimise exposure to air pollution [12, 49]. One example is to facilitate community interventions that aim to reduce the mental health impacts of air pollution while also promoting time outdoors and in green spaces (e.g., walking groups that guide people through places with minimal air pollution). Ideally, community sporting or marathon events could also be held on days or in places where levels of air pollution are minimised [77].

Additionally, communities must ensure that the physical environment supports physical activity and is specifically designed for people [31]. Designing spaces such as greenways and pedestrian areas that are situated away from traffic can improve public health. These strategies not only help to alleviate overcrowding and stress but also highly encourage physical exercise while reducing exposure to air pollutants

[97, 111]. In this context, mental health impacts from overcrowding can be mitigated by creating well-designed urban spaces, reducing congestion, and promoting nature-based solutions. Land-use planning plays a critical role in improving residents' health by incorporating air quality considerations. For instance, developing high-density, mixed-use areas encourages walking and cycling, which reduces emissions while encouraging physical activity [97].

8.7.2.3 Education and Individualised Support

Communities have unique risks and needs as it relates to air pollution, and members living within a given community may benefit from shared discussion and education about ways to protect themselves from its effects. For example, educating residents on how to improve their indoor environment through practices like proper cleaning, ventilation, and pest control can reduce some of the risks associated with air pollution and overcrowding [54]. At the individual level, behaviour change interventions could be strategically developed with healthcare providers and their patients. In particular, monitoring and reducing air pollution exposure have been shown to improve self-reported asthma health and overall quality of life [45, 59]. Minimising exposure to indoor air pollutants can be achieved by reducing pollution sources or changing daily routines. These include increasing or changing the current ventilation routine in the home, using air purifiers or filters, and substituting frequent cooking behaviours for alternative behaviours such as opting to use slow cooking instead of a gas stovetop [59]. Other health protection strategies include staying indoors during high pollution periods, using indoor air filters, wearing face masks, and limiting physical activity when pollution levels are elevated [49].

Not all people will have access to the means, ability, or resources to effectively reduce their exposure to air pollution. Increasing access to mental health support is key when air pollution exposure cannot be directly controlled, and support must be tailored for people with differing psychological needs [11, 20, 98]. Self-help resources, online interventions, or one-to-one intervention with a mental health professional are just some of the ways in which to help protect the well-being of people experiencing poor mental health. Of further consideration is how mental health interventions can be combined with nature or community-based solutions in order to simultaneously support mental health and mitigate the effects of air pollution. For example, community gardening, nature-based therapy, mindfulness in nature, or other nature-based community projects can help to build social support, help people regulate their emotions, and increase contact with green spaces, all whilst reducing exposure to air pollution. Furthermore, collaboration between mental health professionals and other primary care practitioners is particularly important when improving the physical and mental health of people exposed to air pollutants, such as those with a respiratory disease [17].

8.8 Conclusion

This book chapter underscores the growing body of evidence linking air pollution to both physical and mental health risks. While the harmful effects of air pollutants on physical health are well-documented, their impact on mental well-being is gaining increasing attention. Pollutants such as particulate matter, nitrogen dioxide, and sulphur dioxide have been associated with a range of mental health issues, including depression, anxiety, psychosis, and suicidal ideation. These effects are particularly severe in vulnerable populations like pregnant women and fetuses, children and adolescents, the elderly, those with pre-existing health conditions, and lower socioeconomic groups. Understanding the pathways through which pollution influences mental health, whether through direct biological mechanisms like neuroinflammation or indirect factors such as stress and disrupted sleep patterns, is essential for developing targeted interventions. Additionally, addressing broader social determinants, such as housing and poverty, overcrowding, and limited access to green spaces, is crucial in mitigating the mental health burden of air pollution. The broader effects of climate change further exacerbate the threats posed by air pollution. Climate-driven events such as heatwaves and wildfires deteriorate air quality, increasing exposure to harmful pollutants like particulate matter and ozone. Given the complexity of air pollution impacts on mental health, a multidisciplinary approach is necessary. This should include combining public policy, community interventions, and healthcare strategies. By improving air quality, promoting access to green spaces, and providing tailored mental health support, we can protect and enhance the mental well-being of individuals and communities worldwide.

References

1. Ahmed, S. M., Mishra, G. D., Moss, K. M., Yang, I. A., Lycett, K., & Knibbs, L. D. (2022). Maternal and childhood ambient air pollution exposure and mental health symptoms and psychomotor development in children: An Australian population-based longitudinal study. *Environment International, 158*, 107003. https://doi.org/10.1016/j.envint.2021.107003
2. American Psychiatric Association. (2022). *Diagnostic and statistical manual of mental disorders: Text revision (fifth edition)*. American Psychiatric Association. https://doi.org/10.1176/appi.books.9780890425596
3. Bakolis, I., Hammoud, R., Stewart, R., Beevers, S., Dajnak, D., MacCrimmon, S., Broadbent, M., Pritchard, M., Shiode, N., Fecht, D., Gulliver, J., Hotopf, M., Hatch, S. L., & Mudway, I. S. (2021). Mental health consequences of urban air pollution: Prospective population-based longitudinal survey. *Social Psychiatry and Psychiatric Epidemiology, 56*(9), 1587–1599. https://doi.org/10.1007/s00127-020-01966-x
4. Banwell, N., Michel, S., & Senn, N. (2024). Greenspaces and health: Scoping review of studies in Europe. *Public Health Reviews, 45*, 1606863. https://doi.org/10.3389/phrs.2024.1606863
5. Bhui, K., Newbury, J. B., Latham, R. M., Ucci, M., Nasir, Z. A., Turner, B., O'Leary, C., Fisher, H. L., Marczylo, E., Douglas, P., Stansfeld, S., Jackson, S. K., Tyrrel, S., Rzhetsky, A., Kinnersley, R., Kumar, P., Duchaine, C., & Coulon, F. (2023). Air quality and mental

health: Evidence, challenges and future directions. *BJPsych Open, 9*(4), e120. https://doi.org/10.1192/bjo.2023.507

6. Braubach, M., Egorov, A., Mudu, P., Wolf, T., Ward Thompson, C., & Martuzzi, M. (2017). Effects of urban green space on environmental health, equity and resilience. In N. Kabisch, H. Korn, J. Stadler, & A. Bonn (Eds.), *Nature-based solutions to climate change adaptation in urban areas* (pp. 187–205). Springer. https://doi.org/10.1007/978-3-319-56091-5_11
7. Brusselaers, N., Macharis, C., & Mommens, K. (2023). Rerouting urban construction transport flows to avoid air pollution hotspots. *Transportation Research Part D: Transport and Environment, 119*, 103747. https://doi.org/10.1016/j.trd.2023.103747
8. Bruyneel, L., Kestens, W., Alberty, M., Karakaya, G., Van Woensel, R., Horemans, C., Trimpeneers, E., Vanpoucke, C., Fierens, F., Nawrot, T. S., & Cox, B. (2022). Short-term exposure to ambient air pollution and onset of work incapacity related to mental health conditions. *Environment International, 164*, 107245. https://doi.org/10.1016/j.envint.2022.107245
9. Buoli, M., Grassi, S., Caldiroli, A., Carnevali, G. S., Mucci, F., Iodice, S., Cantone, L., Pergoli, L., & Bollati, V. (2018). Is there a link between air pollution and mental disorders? *Environment International, 118*, 154–168. https://doi.org/10.1016/j.envint.2018.05.044
10. Campbell, C. E., Cotter, D. L., Bottenhorn, K. L., Burnor, E., Ahmadi, H., Gauderman, W. J., Cardenas-Iniguez, C., Hackman, D., McConnell, R., Berhane, K., Schwartz, J., Chen, J.-C., & Herting, M. M. (2024). Air pollution and age-dependent changes in emotional behavior across early adolescence in the U.S. *Environmental Research, 240*(Pt 1), 117390. https://doi.org/10.1016/j.envres.2023.117390
11. Cao, Z., Zhou, J., Li, M., Huang, J., & Dou, D. (2023). Urbanites' mental health undermined by air pollution. *Nature Sustainability, 6*(4), 470–478. https://doi.org/10.1038/s41893-022-01032-1
12. Carlisle, A. J., & Sharp, N. C. C. (2001). Exercise and outdoor ambient air pollution. *British Journal of Sports Medicine, 35*(4), 214–222. https://doi.org/10.1136/bjsm.35.4.214
13. Carlsten, C., Salvi, S., Wong, G. W. K., & Chung, K. F. (2020). Personal strategies to minimise effects of air pollution on respiratory health: Advice for providers, patients and the public. *European Respiratory Journal, 55*(6), 1902056. https://doi.org/10.1183/13993003.02056-2019
14. Chen, S., Kong, J., Yu, F., & Peng, K. (2018). Psychopathological symptoms under smog: The role of emotion regulation. *Frontiers in Psychology, 8*, Article 2274. https://doi.org/10.3389/fpsyg.2017.02274
15. Chen, S., Oliva, P., & Zhang, P. (2024). Air pollution and mental health: Evidence from China. *AEA Papers and Proceedings, 114*, 423–428. https://doi.org/10.1257/pandp.20241062
16. Choi, K.-H., Bae, S., Kim, S., & Kwon, H.-J. (2020). Indoor and outdoor PM2.5 exposure, and anxiety among schoolchildren in Korea: A panel study. *Environmental Science and Pollution Research International, 27*(22), 27984–27994. https://doi.org/10.1007/s11356-020-08900-3
17. Cientanni, F., Liu, P., Galliard, H., Fearn, J., & Dhasmana, D. (2023). P30 mental health interventions in chronic respiratory disease (mind project): A pilot trial of specialised psychological care embedded within a general respiratory service. *Thorax, 78*(Suppl 4), A124–A124. https://doi.org/10.1136/thorax-2023-BTSabstracts.182
18. Clark, D. P. Q., Son, D. B., Bowatte, G., Senaratna, C. V., Lodge, C., Perret, J. L., Hamilton, G. S., & Dharmage, S. (2020). The association between traffic-related air pollution and obstructive sleep apnea: A systematic review. *Sleep Medicine Reviews, 54*, 101360. https://doi.org/10.1016/j.smrv.2020.101360
19. Clifford, A., Lang, L., Chen, R., Anstey, K. J., & Seaton, A. (2016). Exposure to air pollution and cognitive functioning across the life course—A systematic literature review. *Environmental Research, 147*, 383–398. https://doi.org/10.1016/j.envres.2016.01.018
20. Cuijpers, P., Miguel, C., Ciharova, M., Kumar, M., Brander, L., Kumar, P., & Karyotaki, E. (2023). Impact of climate events, pollution, and green spaces on mental health: An umbrella review of meta-analyses. *Psychological Medicine, 53*(3), 638–653. https://doi.org/10.1017/S0033291722003890

21. Davis, E. P., & Narayan, A. J. (2020). Pregnancy as a period of risk, adaptation, and resilience for mothers and infants. *Development and Psychopathology, 32*(5), 1625–1639. https://doi.org/10.1017/S0954579420001121
22. deSouza, P. N., Anenberg, S., Fann, N., McKenzie, L. M., Chan, E., Roy, A., Jimenez, J. L., Raich, W., Roman, H., & Kinney, P. L. (2024). Evaluating the sensitivity of mortality attributable to pollution to modeling choices: A case study for Colorado. *Environment International, 185*, 108416. https://doi.org/10.1016/j.envint.2024.108416
23. deSouza, P. N., Chaudhary, E., Dey, S., Ko, S., Németh, J., Guttikunda, S., Chowdhury, S., Kinney, P., Subramanian, S. V., Bell, M. L., & Kim, R. (2023). An environmental justice analysis of air pollution in India. *Scientific Reports, 13*(1), 16690. https://doi.org/10.1038/s41598-023-43628-3
24. Du, J., Cui, Y., Yang, L., Duan, Y., Qi, Q., & Liu, H. (2024). Associations of indoor ventilation frequency with depression and anxiety in chinese older adults. *Indoor Air, 2024*, 1–11. https://doi.org/10.1155/2024/9943687
25. Elavsky, S., Jandačková, V., Knapová, L., Vašendová, V., Sebera, M., Kaštovská, B., Blaschová, D., Kühnová, J., Cimler, R., Vilímek, D., Bosek, T., Koenig, J., & Jandačka, D. (2021). Physical activity in an air-polluted environment: Behavioral, psychological and neuroimaging protocol for a prospective cohort study (Healthy aging in industrial environment study—Program 4). *BMC Public Health, 21*(1), 126. https://doi.org/10.1186/s12889-021-10166-4
26. European Environment Agency. (2019). *Healthy environment, healthy lives: How the environment influences health and well-being in Europe*. https://www.eea.europa.eu/publications/healthy-environment-healthy-lives
27. Firdaus, G. (2018). Increasing rate of psychological distress in urban households: How does income matter? *Community Mental Health Journal, 54*(5), 641–648. https://doi.org/10.1007/s10597-017-0193-9
28. Firdaus, G., & Ahmad, A. (2013). Relationship between Housing and Health: A cross-sectional study of an urban centre of India. *Indoor and Built Environment, 22*(3), 498–507. https://doi.org/10.1177/1420326X12443846
29. Gao, Q., Xu, Q., Guo, X., Fan, H., & Zhu, H. (2017). Particulate matter air pollution associated with hospital admissions for mental disorders: A time-series study in Beijing, China. *European Psychiatry: The Journal of the Association of European Psychiatrists, 44*, 68–75. https://doi.org/10.1016/j.eurpsy.2017.02.492
30. Geneshka, M., Coventry, P., Cruz, J., & Gilbody, S. (2021). Relationship between green and blue spaces with mental and physical health: A systematic review of longitudinal observational studies. *International Journal of Environmental Research and Public Health, 18*(17), 9010. https://doi.org/10.3390/ijerph18179010
31. Giles, L. V., Koehle, M. S., Saelens, B. E., Sbihi, H., & Carlsten, C. (2021). When physical activity meets the physical environment: Precision health insights from the intersection. *Environmental Health and Preventive Medicine, 26*(1), 68. https://doi.org/10.1186/s12199-021-00990-w
32. Glinianaia, S. V., Rankin, J., Bell, R., Pless-Mulloli, T., & Howel, D. (2004). Does particulate air pollution contribute to infant death? A systematic review. *Environmental Health Perspectives, 112*(14), 1365–1370. https://doi.org/10.1289/ehp.6857
33. Gumy, S., & Mudu, P. (2018). News on air pollution and health data and impacts on health from the World Health Organization. *Clean Air Journal*. https://doi.org/10.17159/2410-972X/2018/v28n1a2
34. Hartinger, S. M., Palmeiro-Silva, Y. K., Llerena-Cayo, C., Blanco-Villafuerte, L., Escobar, L. E., Diaz, A., Sarmiento, J. H., Lescano, A. G., Melo, O., Rojas-Rueda, D., Takahashi, B., Callaghan, M., Chesini, F., Dasgupta, S., Posse, C. G., Gouveia, N., Martins De Carvalho, A., Miranda-Chacón, Z., Mohajeri, N., et al. (2024). The 2023 Latin America report of the Lancet Countdown on health and climate change: The imperative for health-centred climate-resilient

development. *The Lancet Regional Health Americas, 33*, 100746. https://doi.org/10.1016/j.lana.2024.100746

35. Hautekiet, P., Saenen, N. D., Demarest, S., Keune, H., Pelgrims, I., Van der Heyden, J., De Clercq, E. M., & Nawrot, T. S. (2022). Air pollution in association with mental and self-rated health and the mediating effect of physical activity. *Environmental Health, 21*(1), 29. https://doi.org/10.1186/s12940-022-00839-x
36. Health Effects Institute. (2024). *State of Global AIR/2024 [Special report]*. Health Effects Institute. https://www.stateofglobalair.org/resources/report/state-global-air-report-2024
37. Holden, K. A., Lee, A. R., Hawcutt, D. B., & Sinha, I. P. (2023). The impact of poor housing and indoor air quality on respiratory health in children. *Breathe, 19*(2), 230058. https://doi.org/10.1183/20734735.0058-2023
38. Hong, J., Kang, J. M., Cho, S.-E., Jung, J., & Kang, S.-G. (2023). Significant association between increased risk of emergency department visits for psychiatric disorders and air pollutants in South Korea. *Journal of Exposure Science & Environmental Epidemiology, 33*(3), 490–499. https://doi.org/10.1038/s41370-022-00504-y
39. Hou, Y., Gao, M., Huang, L., & Wang, Q. (2021). *Air pollution reduces interpersonal trust: The roles of emotion and emotional susceptibility*. https://www.mdpi.com/1660-4601/18/11/5631
40. Humphreys, A., Walker, E. G., Bratman, G. N., & Errett, N. A. (2022). What can we do when the smoke rolls in? An exploratory qualitative analysis of the impacts of rural wildfire smoke on mental health and wellbeing, and opportunities for adaptation. *BMC Public Health, 22*(1), 41. https://doi.org/10.1186/s12889-021-12411-2
41. Hyslop, N. P. (2009). Impaired visibility: The air pollution people see. *Atmospheric Environment, 43*(1), 182–195. https://doi.org/10.1016/j.atmosenv.2008.09.067
42. Jbaily, A., Zhou, X., Liu, J., Lee, T.-H., Kamareddine, L., Verguet, S., & Dominici, F. (2022). Air pollution exposure disparities across US population and income groups. *Nature, 601*(7892), 228–233. https://doi.org/10.1038/s41586-021-04190-y
43. Jorcano, A., Lubczyńska, M. J., Pierotti, L., Altug, H., Ballester, F., Cesaroni, G., El Marroun, H., Fernández-Somoano, A., Freire, C., Hanke, W., Hoek, G., Ibarluzea, J., Iñiguez, C., Jansen, P. W., Lepeule, J., Markevych, I., Polańska, K., Porta, D., Schikowski, T., et al. (2019). Prenatal and postnatal exposure to air pollution and emotional and aggressive symptoms in children from 8 European birth cohorts. *Environment International, 131*, 104927. https://doi.org/10.1016/j.envint.2019.104927
44. Kajjoba, D., Kasedde, H., Kirabira, J. B., Wesonga, R., Mugwanya, R., Lwanyaga, J. D., & Olupot, P. W. (2024). Impact of natural ventilation and outdoor environment on indoor air quality and occupant health in low-income tropical housing. *Energy Reports, 12*, 4184–4194. https://doi.org/10.1016/j.egyr.2024.10.004
45. Kang, J., Kim, H.-C., Jang, Y., Lee, J. B., Lee, J. S., Oh, Y.-M., Ji, H. W., Jung, J. Y., & Lee, S. W. (2023). Randomised controlled trial of a behavioural intervention to reduce exposure to PM2.5 in patients with COPD. *Environment International, 181*, 108286. https://doi.org/10.1016/j.envint.2023.108286
46. Kolokotroni, M., Ge, Y. T., & Katsoulas, D. (2002). Monitoring and modelling indoor air quality and ventilation in classrooms within a purpose-designed naturally ventilated school. *Indoor and Built Environment, 11*(6), 316–326. https://doi.org/10.1177/1420326X0201100603
47. Kumar, N., Kumar, S., & Pandey, S. P. (2023). Traffic-related air pollution and associated human health risk. *Macromolecular Symposia, 407*(1), 2100486. https://doi.org/10.1002/masy.202100486
48. Lai, W., Li, S., Li, Y., & Tian, X. (2022). Air pollution and cognitive functions: Evidence from straw burning in China. *American Journal of Agricultural Economics, 104*(1), 190–208. https://doi.org/10.1111/ajae.12225
49. Laumbach, R. J., & Cromar, K. R. (2022). Personal interventions to reduce exposure to outdoor air pollution. *Annual Review of Public Health, 43*(1), 293–309. https://doi.org/10.1146/annurev-publhealth-052120-103607

50. Li, X., & Tilt, B. (2018). Perceptions of quality of life and pollution among China's urban middle class: The case of smog in Tangshan. *The China Quarterly, 234*, 340–356. https://doi.org/10.1017/S0305741017001382
51. Li, Y., Guan, D., Tao, S., Wang, X., & He, K. (2018). A review of air pollution impact on subjective well-being: Survey versus visual psychophysics. *Journal of Cleaner Production, 184*, 959–968. https://doi.org/10.1016/j.jclepro.2018.02.296
52. Lim, Y.-H., Kim, H., Kim, J. H., Bae, S., Park, H. Y., & Hong, Y.-C. (2012). Air pollution and symptoms of depression in elderly adults. *Environmental Health Perspectives, 120*(7), 1023–1028. https://doi.org/10.1289/ehp.1104100
53. Liu, J., Wu, T., Liu, Q., Wu, S., & Chen, J.-C. (2020). Air pollution exposure and adverse sleep health across the life course: A systematic review. *Environmental Pollution, 262*, 114263. https://doi.org/10.1016/j.envpol.2020.114263
54. Lorentzen, J. C., Johanson, G., Björk, F., & Stensson, S. (2022). Overcrowding and hazardous dwelling condition characteristics: A Systematic search and scoping review of relevance for health. *International Journal of Environmental Research and Public Health, 19*(23), 15542. https://doi.org/10.3390/ijerph192315542
55. Makri, A., & Stilianakis, N. I. (2008). Vulnerability to air pollution health effects. *International Journal of Hygiene and Environmental Health, 211*(3), 326–336. https://doi.org/10.1016/j.ijheh.2007.06.005
56. Manisalidis, I., Stavropoulou, E., Stavropoulos, A., & Bezirtzoglou, E. (2020). Environmental and health impacts of air pollution: A review. *Frontiers in Public Health, 8*, 14. https://doi.org/10.3389/fpubh.2020.00014
57. Marazziti, D., Cianconi, P., Mucci, F., Foresi, L., Chiarantini, I., & Della Vecchia, A. (2021). Climate change, environment pollution, COVID-19 pandemic and mental health. *Science of the Total Environment, 773*, 145182. https://doi.org/10.1016/j.scitotenv.2021.145182
58. Marmett, B., Carvalho, R. B., Dorneles, G. P., Nunes, R. B., & Rhoden, C. R. (2020). Should I stay or should I go: Can air pollution reduce the health benefits of physical exercise? *Medical Hypotheses, 144*, 109993. https://doi.org/10.1016/j.mehy.2020.109993
59. McCarron, A., Semple, S., Swanson, V., Gillespie, C., Braban, C., & Price, H. D. (2024). Piloting co-developed behaviour change interventions to reduce exposure to air pollution and improve self-reported asthma-related health. *Journal of Exposure Science & Environmental Epidemiology*. https://doi.org/10.1038/s41370-024-00661-2
60. Milner, J., Davies, M., & Wilkinson, P. (2012). Urban energy, carbon management (low carbon cities) and co-benefits for human health. *Current Opinion in Environmental Sustainability, 4*(4), 398–404. https://doi.org/10.1016/j.cosust.2012.09.011
61. Mota-Bertran, A., Coenders, G., Plaja, P., Saez, M., & Barceló, M. A. (2024). Air pollution and children's mental health in rural areas: Compositional spatio-temporal model. *Scientific Reports, 14*(1), 19363. https://doi.org/10.1038/s41598-024-70024-2
62. Muhsin, H. A., Steingrimsson, S., Oudin, A., Åström, D. O., & Carlsen, H. K. (2022). Air pollution and increased number of psychiatric emergency room visits: A case-crossover study for identifying susceptible groups. *Environmental Research, 204*, 112001. https://doi.org/10.1016/j.envres.2021.112001
63. Muindi, K., Egondi, T., Kimani-Murage, E., Rocklov, J., & Ng, N. (2014). "We are used to this": A qualitative assessment of the perceptions of and attitudes towards air pollution amongst slum residents in Nairobi. *BMC Public Health, 14*(1), 226. https://doi.org/10.1186/1471-2458-14-226
64. Munroe, R. L., & Gauvain, M. (2012). Exposure to open-fire cooking and cognitive performance in children. *International Journal of Environmental Health Research, 22*(2), 156–164. https://doi.org/10.1080/09603123.2011.628642
65. National Institute of Environmental Health Sciences. (2024). *Air pollution and your health*. https://www.niehs.nih.gov/health/topics/agents/air-pollution
66. Newbury, J. B., Arseneault, L., Beevers, S., Kitwiroon, N., Roberts, S., Pariante, C. M., Kelly, F. J., & Fisher, H. L. (2019). Association of air pollution exposure with psychotic

experiences during adolescence. *JAMA Psychiatry, 76*(6), 614–623. https://doi.org/10.1001/jamapsychiatry.2019.0056

67. Newbury, J. B., Stewart, R., Fisher, H. L., Beevers, S., Dajnak, D., Broadbent, M., Pritchard, M., Shiode, N., Heslin, M., Hammoud, R., Hotopf, M., Hatch, S. L., Mudway, I. S., & Bakolis, I. (2021). Association between air pollution exposure and mental health service use among individuals with first presentations of psychotic and mood disorders: Retrospective cohort study. *The British Journal of Psychiatry, 219*(6), 678–685. https://doi.org/10.1192/bjp.2021.119
68. Nguyen, A.-M., Malig, B. J., & Basu, R. (2021). The association between ozone and fine particles and mental health-related emergency department visits in California, 2005–2013. *PLoS One, 16*(4), e0249675. https://doi.org/10.1371/journal.pone.0249675
69. Noël, C., Van Landschoot, L., Vanroelen, C., & Gadeyne, S. (2022). The public's perceptions of air pollution. What's in a name? *Environmental Health Insights, 16*, 11786302221123563. https://doi.org/10.1177/11786302221123563
70. Palagini, L., Hertenstein, E., Riemann, D., & Nissen, C. (2022). Sleep, insomnia and mental health. *Journal of Sleep Research, 31*(4), e13628. https://doi.org/10.1111/jsr.13628
71. Patel, L., Friedman, E., Johannes, S. A., Lee, S. S., O'Brien, H. G., & Schear, S. E. (2021). Air pollution as a social and structural determinant of health. *The Journal of Climate Change and Health, 3*, 100035. https://doi.org/10.1016/j.joclim.2021.100035
72. Perera, F., Ashrafi, A., Kinney, P., & Mills, D. (2019). Towards a fuller assessment of benefits to children's health of reducing air pollution and mitigating climate change due to fossil fuel combustion. *Environmental Research, 172*, 55–72. https://doi.org/10.1016/j.envres.2018.12.016
73. Petrowski, K., Bührer, S., Strauß, B., Decker, O., & Brähler, E. (2021). Examining air pollution (PM10), mental health and well-being in a representative German sample. *Scientific Reports, 11*(1), 18436. https://doi.org/10.1038/s41598-021-93773-w
74. Pignon, B., Borel, C., Lajnef, M., Richard, J.-R., Szöke, A., Hemery, F., Leboyer, M., Foret, G., & Schürhoff, F. (2022). PM2.5 and PM10 air pollution peaks are associated with emergency department visits for psychotic and mood disorders. *Environmental Science and Pollution Research, 29*(59), 88577–88586. https://doi.org/10.1007/s11356-022-21964-7
75. Pinho-Gomes, A.-C., Roaf, E., Fuller, G., Fowler, D., Lewis, A., ApSimon, H., Noakes, C., Johnstone, P., & Holgate, S. (2023). Air pollution and climate change. *The Lancet Planetary Health, 7*(9), e727–e728. https://doi.org/10.1016/S2542-5196(23)00189-4
76. Pun, V. C., Manjourides, J., & Suh, H. (2017). Association of ambient air pollution with depressive and anxiety symptoms in older adults: Results from the NSHAP study. *Environmental Health Perspectives, 125*(3), 342–348. https://doi.org/10.1289/EHP494
77. Qin, F., Yang, Y., Wang, S., Dong, Y., Xu, M., Wang, Z., & Zhao, J. (2019). Exercise and air pollutants exposure: A systematic review and meta-analysis. *Life Sciences, 218*, 153–164. https://doi.org/10.1016/j.lfs.2018.12.036
78. Qiu, X., Shi, L., Kubzansky, L. D., Wei, Y., Castro, E., Li, H., Weisskopf, M. G., & Schwartz, J. D. (2023). Association of long-term exposure to air pollution with late-life depression in older adults in the US. *JAMA Network Open, 6*(2), e2253668. https://doi.org/10.1001/jamanetworkopen.2022.53668
79. Ramar, K., Malhotra, R. K., Carden, K. A., Martin, J. L., Abbasi, F. F., Aurora, R. N., Kapur, V. K., Olson, E. J., Rosen, C. L., Rowley, J. A., Shelgikar, A. V., & Trotti, L. M. (n.d.). Sleep is essential to health: An American Academy of Sleep Medicine position statement. *Journal of Clinical Sleep Medicine, 17*(10), 2115–2119. https://doi.org/10.5664/jcsm.9476
80. Reeve, I., Scott, J., Hine, D. W., & Bhullar, N. (2013). "This is not a burning issue for me": How citizens justify their use of wood heaters in a city with a severe air pollution problem. *Energy Policy, 57*, 204–211. https://doi.org/10.1016/j.enpol.2013.01.042
81. Reuben, A., Manczak, E. M., Cabrera, L. Y., Alegria, M., Bucher, M. L., Freeman, E. C., Miller, G. W., Solomon, G. M., & Perry, M. J. (2022). The Interplay of environmental

exposures and mental health: Setting an agenda. *Environmental Health Perspectives, 130*(2), 025001. https://doi.org/10.1289/EHP9889
82. Roberts, S., Arseneault, L., Barratt, B., Beevers, S., Danese, A., Odgers, C. L., Moffitt, T. E., Reuben, A., Kelly, F. J., & Fisher, H. L. (2019). Exploration of NO2 and PM2.5 air pollution and mental health problems using high-resolution data in London-based children from a UK longitudinal cohort study. *Psychiatry Research, 272*, 8–17. https://doi.org/10.1016/j.psychres.2018.12.050
83. Roe, J., Mondschein, A., Neale, C., Barnes, L., Boukhechba, M., & Lopez, S. (2020). The urban built environment, walking and mental health outcomes among older adults: A pilot study. *Frontiers in Public Health, 8*, 575946. https://doi.org/10.3389/fpubh.2020.575946
84. Romanova, E. (2018). Increase in population density and aggravation of social and psychological problems in areas with high-rise construction. *E3S Web of Conferences, 33*, 03061. https://doi.org/10.1051/e3sconf/20183303061
85. Ronaldson, A., Stewart, R., Mueller, C., Das-Munshi, J., Newbury, J. B., Mudway, I. S., Broadbent, M., Fisher, H. L., Beevers, S., Dajnak, D., Hotopf, M., Hatch, S. L., & Bakolis, I. (2023). Associations between air pollution and mental health service use in dementia: A retrospective cohort study. *BMJ Ment Health, 26*(1), e300762. https://doi.org/10.1136/bmjment-2023-300762
86. Ruiz-Tagle, J., & Urria, I. (2022). Household overcrowding trajectories and mental well-being. *Social Science & Medicine, 296*, 114051. https://doi.org/10.1016/j.socscimed.2021.114051
87. Saadeh, R., Khader, Y., Malkawi, M., & Allouh, M. Z. (2022). Communicating the risks of air pollution to the public: A perspective from Jordan and Lebanon. *Environmental Health Insights, 16*, 11786302221127851. https://doi.org/10.1177/11786302221127851
88. Scott, A. J., Webb, T. L., Martyn-St James, M., Rowse, G., & Weich, S. (2021). Improving sleep quality leads to better mental health: A meta-analysis of randomised controlled trials. *Sleep Medicine Reviews, 60*, 101556. https://doi.org/10.1016/j.smrv.2021.101556
89. Solari, C. D., & Mare, R. D. (2012). Housing crowding effects on children's wellbeing. *Social Science Research, 41*(2), 464–476. https://doi.org/10.1016/j.ssresearch.2011.09.012
90. Song, R., Liu, L., Wei, N., Li, X., Liu, J., Yuan, J., Yan, S., Sun, X., Mei, L., Liang, Y., Li, Y., Jin, X., Wu, Y., Pan, R., Yi, W., Song, J., He, Y., Tang, C., Liu, X., et al. (2023). Short-term exposure to air pollution is an emerging but neglected risk factor for schizophrenia: A systematic review and meta-analysis. *Science of the Total Environment, 854*, 158823. https://doi.org/10.1016/j.scitotenv.2022.158823
91. Surace, T., Quitadamo, C., Caldiroli, A., Capuzzi, E., Colmegna, Nosari, G., Borroni, E., Fedrizzi, L., Bollati, V., Pesatori, A. C., Carugno, M., Clerici, M., & Buoli, M. (2023). Air pollution and perinatal mental health: A comprehensive overview. *Journal of Clinical Medicine, 12*(9) https://www.mdpi.com/2077-0383/12/9/3146
92. Theron, L. C., Abreu-Villaça, Y., Augusto-Oliveira, M., Brennan, C., Crespo-Lopez, M. E., de Paula Arrifano, G., Glazer, L., Gwata, N., Lin, L., Mareschal, I., Mermelstein, S., Sartori, L., Stieger, L., Trotta, A., & Hadfield, K. (2022). A systematic review of the mental health risks and resilience among pollution-exposed adolescents. *Journal of Psychiatric Research, 146*, 55–66. https://doi.org/10.1016/j.jpsychires.2021.12.012
93. Thurston, G. D., & Bell, M. L. (2014). The human health co-benefits of air quality improvements associated with climate change mitigation. In K. E. Pinkerton & W. N. Rom (Eds.), *Global climate change and public health* (pp. 137–154). Springer. https://doi.org/10.1007/978-1-4614-8417-2_8
94. Vardoulakis, S., & Osborne, N. (2018). Air pollution and asthma. *Archives of Disease in Childhood, 103*(9), 813–814. https://doi.org/10.1136/archdischild-2017-314543
95. Vert, C., Sánchez-Benavides, G., Martínez, D., Gotsens, X., Gramunt, N., Cirach, M., Molinuevo, J. L., Sunyer, J., Nieuwenhuijsen, M. J., Crous-Bou, M., & Gascon, M. (2017). Effect of long-term exposure to air pollution on anxiety and depression in adults: A cross-sectional study. *International Journal of Hygiene and Environmental Health, 220*(6), 1074–1080. https://doi.org/10.1016/j.ijheh.2017.06.009

96. Vig, P. S., Lim, J. Y., Lee, R. W. L., Huang, H., Tan, X. H., Lim, W. Q., Lim, M. B. X. Y., Lee, A. S. I., Chiam, M., Lim, C., Baral, V. R., & Krishna, L. K. R. (2021). Parental bereavement—Impact of death of neonates and children under 12 years on personhood of parents: A systematic scoping review. *BMC Palliative Care, 20*(1), 136. https://doi.org/10.1186/s12904-021-00831-1
97. Washam, C. (2011). Separating people from pollution: Individual and community interventions to mitigate health effects of air pollutants. *Environmental Health Perspectives, 119*(1), a. https://doi.org/10.1289/ehp.119-a34a
98. Werder, E., Lawrence, K., Deng, X., Braxton Jackson, W., Christenbury, K., Buller, I., Engel, L., & Sandler, D. (2024). Residential air pollution, greenspace, and adverse mental health outcomes in the U.S. Gulf Long-term Follow-up Study. *Science of the Total Environment, 946*, 174434. https://doi.org/10.1016/j.scitotenv.2024.174434
99. World Health Organization. (n.d.-a). *Air pollution*. Retrieved October 1, 2024, from https://www.who.int/health-topics/air-pollution
100. World Health Organization. (n.d.-b). *Air pollution and health training toolkit for health workers (APHT)*. Retrieved October 22, 2024, from https://www.who.int/tools/air-pollution-and-health-training-toolkit-for-health-workers
101. World Health Organization. (n.d.-c). *Strategies for reducing exposure to household air pollution include clean fuels and stoves, supportive policies and awareness-raising campaigns*. Retrieved October 22, 2024, from https://www.who.int/teams/environment-climate-change-and-health/air-quality-energy-and-health/sectoral-interventions/household-air-pollution/strategies
102. World Health Organization. (2017). *Urban green spaces: A brief for action*. https://www.who.int/europe/publications/i/item/9789289052498
103. World Health Organization. (2023). *Climate change*. https://www.who.int/news-room/fact-sheets/detail/climate-change-and-health
104. Xie, Y., Dai, H., Xu, X., Fujimori, S., Hasegawa, T., Yi, K., Masui, T., & Kurata, G. (2018). Co-benefits of climate mitigation on air quality and human health in Asian countries. *Environment International, 119*, 309–318. https://doi.org/10.1016/j.envint.2018.07.008
105. Xue, T., Guan, T., Zheng, Y., Geng, G., Zhang, Q., Yao, Y., & Zhu, T. (2021). Long-term PM2.5 exposure and depressive symptoms in China: A quasi-experimental study. *The Lancet Regional Health Western Pacific, 6*, 100079. https://doi.org/10.1016/j.lanwpc.2020.100079
106. Yolton, K., Khoury, J. C., Burkle, J., LeMasters, G., Cecil, K., & Ryan, P. (2019). Lifetime exposure to traffic-related air pollution and symptoms of depression and anxiety at age 12 years. *Environmental Research, 173*, 199–206.
107. Yoo, E., Roberts, J. E., & Chen, K. (2024). Effects of air pollution on emergency room visits for mental disorders: Risks and effect modification by comorbid physical disorders and personal characteristics. *Environmental Research: Health, 2*(2), 025001. https://doi.org/10.1088/2752-5309/ad1f3f
108. Yoo, E.-H., Roberts, J. E., Eum, Y., Li, X., Chu, L., Wang, P., & Chen, K. (2022). Short-term exposure to air pollution and mental disorders: A case-crossover study in New York City. *Environmental Research: Health, 1*(1), 015001. https://doi.org/10.1088/2752-5309/ac6439
109. Yu, W., & Weitzman, M. (2022). Climate change, air pollution, and children's mental health. *Journal of Applied Research on Children: Informing Policy for Children at Risk, 12*(1). https://doi.org/10.58464/2155-5834.1491
110. Zhang, Q., Meng, X., Luo, H., Yu, K., Li, A., Zhou, L., Chen, R., & Kan, H. (2024). Air pollutants, genetic susceptibility, and incident schizophrenia in later life: A prospective study in the UK Biobank. *Science of the Total Environment, 934*, 173009. https://doi.org/10.1016/j.scitotenv.2024.173009
111. Zhang, Z., Měchurová, K., Resch, B., Amegbor, P., & Sabel, C. E. (2023). Assessing the association between overcrowding and human physiological stress response in different urban contexts: A case study in Salzburg, Austria. *International Journal of Health Geographics, 22*(1), 15. https://doi.org/10.1186/s12942-023-00334-7

112. Zhou, Y., & Liu, J. (2020). Air pollution and mental health of older adults in China. *Sustainability, 12*(3), Article 3. https://doi.org/10.3390/su12030950
113. Zundel, C. G., Ryan, P., Brokamp, C., Heeter, A., Huang, Y., Strawn, J. R., & Marusak, H. A. (2022). Air pollution, depressive and anxiety disorders, and brain effects: A systematic review. *Neurotoxicology, 93*, 272–300. https://doi.org/10.1016/j.neuro.2022.10.011
114. Kim, Y., Manley, J. & Radoias, V. (2020). Air pollution and long term mental health. *Atmosphere, 11*(12), 1355. https://doi.org/10.3390/atmos11121355
115. Borroni, E., Pesatori, A. C., Bollati, V., Buoli, M., & Carugno, C. (2022). Air pollution exposure and depression: A comprehensive updated systematic review and meta-analysis. *Environmental Pollution, 292,* 18245. https://doi.org/10.1016/j.envpol.2021.118245
116. Shin, J., Park, J. Y. & Choi, J. (2018). Long-term exposure to ambient air pollutants and mental health status: A nationwide population-based cross-sectional study. *PLOS ONE, 13*(4), e0195607. https://doi.org/10.1371/journal.pone.0195607
117. Li, H., Zhang, S., Qian, Z., Xie, X-H., Luo, Y., Han, R.,Hou, J., Wang, C., McMillin, S. E., Wu, S., Tian, F., Deng, W-F. & Lin, H. (2020). Short-term effects of air pollution on cause-specific mental disorders in three subtropical Chinese cities. *Environmental Research, 191,* 110214. https://doi.org/10.1016/j.envres.2020.110214
118. Fan, S-J.,Heinric, J., Bloom, M.S., Zhao, T-Y., Shi, T-X., Feng, W-R., Sun, Y., Shen, J-C., Yang, Z-C., Yang, B-Y., Dong, G-H. (2020). Ambient air pollution and depression: A systematic review with meta-analysis up to 2019. *Science of The Total Environment, 701,* 134721. https://doi.org/10.1016/j.scitotenv.2019.134721
119. Gładka, A., Rymaszewska, J. & Zatoński, T. Impact of air pollution on depression and suicide. *International Journal of Occupational Medicine and Environmental Health, 31*(6), 711–721. https://doi.org/10.13075/ijomeh.1896.01277
120. Yang, Z., Song, Q., Li, J., Zhang, Y., Yuan, X-C., Wang, W. & Yu, Q. (2021). Air pollution and mental health: The moderator effect of health behaviors. *Environmental Research Letters, 16*(4), 044005. https://doi.org/10.1088/1748-9326/abe88f
121. Liu, J., Wu, T., Liu, Q., Wu, S. & Chen, J-C. (2020). Air pollution exposure and adverse sleep health across the life course: A systematic review. *Environmental Pollution, 262,* 114263. https://doi.org/10.1016/j.envpol.2020.114263
122. Liu, Q., Wang, W., Gu, X., Deng, F., Wang, X., Lin, H., Gao, X. & Wu, S. (2021). Association between particulate matter air pollution and risk of depression and suicide: a systematic review and meta-analysis. *Environmental Science and Pollution Research, 28*(8), 9029–9049. https://doi.org/10.1007/s11356-021-12357-3
123. Braithwaite, I., Zhang, S., Kirkbride, J. B., Osborn, D. P. J. & Hayes, J. F. (2019). Air pollution (particulate matter) exposure and associations with depression anxiety bipolar psychosis and suicide risk: A systematic review and meta-analysis. *Environmental Health Perspectives, 127*(12), 126002. https://doi.org/10.1289/EHP4595
124. Cao, B., Chen, Y. & McIntyre, R. S. (2021). Comprehensive review of the current literature on impact of ambient air pollution and sleep quality. *Sleep Medicine, 79*, 211–219. https://doi.org/21910.1016/j.sleep.2020.04.009
125. Power, M. C., Kioumourtzoglou, M-A., Hart, J.E., Okereke, O. I., Laden, F. & Weisskopf, M. G. The relation between past exposure to fine particulate air pollution and prevalent anxiety: Observational cohort study. *BMJ, 350,* h1111. https://doi.org/10.1136/bmj.h1111
126. Power, M. C., Adar, S. S., Yanosky, J.D., Weuve, J. (2016). Exposure to air pollution as a potential contributor to cognitive function cognitive decline brain imaging and dementia: A systematic review of epidemiologic research. *NeuroToxicology, 56,* 235. https://doi.org/25310.1016/j.neuro.2016.06.004
127. Kanner, J., Pollack, A. Z., Ranasinghe, S., Stevens, D. R., Nobles, C., Rohn, M. C. H., Sherman, S. & Mendola, P. (2021). Chronic exposure to air pollution and risk of mental health disorders complicating pregnancy. *Environmental Research, 196,* 110937. https://doi.org/10.1016/j.envres.2021.110937

128. Caulfield, J. I. (2021). Anxiety depression and asthma: New perspectives and approaches for psychoneuroimmunology research. *Brain Behavior & Immunity - Health, 18,* 100360. https://doi.org/10.1016/j.bbih.2021.100360
129. Goodwin, R. D., Jacobi, F. & Thefeld, W. (2003). Mental Disorders and Asthma in the Community. *Archives of General Psychiatry, 60*(11), 1125–1130. https://doi.org/10.1001/archpsyc.60.11.1125
130. Cho, J., Choi, Y. J., Suh, S., Sohn, J., Kim, H., Cho, S-K., Ha, K-H., Kim, C., & Shin, D. C. (2014). Air pollution as a risk factor for depressive episode in patients with cardiovascular disease diabetes mellitus or asthma. *Journal of Affective Disorders, 157,* 45–51. https://doi.org/10.1016/j.jad.2014.01.002
131. Hu, Q., Feng, Y. & Xu, M. (2021). Are there heterogeneous impacts of air pollution on mental health? *Frontiers in Public Health,* 9. https://doi.org/10.3389/fpubh.2021.780022
132. Colloca, L. & Barsky, A. J. (2020). Placebo and nocebo effects. *New England Journal of Medicine, 382*(6), 554–561. https://doi.org/10.1056/NEJMra1907805
133. Clayton, S. & Crandon, T. (2025). Climate change and mental health. *Annual Review of Clinical Psychology, 21*(1), 61–83. https://doi.org/10.1146/annurev-clinpsy-081423-025932

Chapter 9
Clearing the Air: Navigating Regulatory Frameworks for Indoor Air Quality Control

Amanda Kennedy, Lidia Morawska, and Vivian Adotey

Keywords Indoor air quality · Public health regulation · Environmental governance · Comparative law · Regulatory innovation

9.1 Introduction

The regulation of indoor air quality (IAQ) has gained increasing prominence worldwide as awareness of the health impacts associated with poor IAQ has grown. Poor IAQ is linked to numerous negative health outcomes, both short-term (acute) and long-term (chronic), including respiratory, cardiovascular, reproductive, and mental health, with health outcomes also including lung cancer, diabetes, leukaemia, and Parkinson's disease [1, 10, 14, 15, 22, 23], as well as infection transmission. Short-term health impacts associated with indoor air pollution include irritation of the eyes, nose, throat, and skin; exacerbation of asthma and allergy symptoms; reduced sleep quality; and neurobehavioral symptoms (e.g., difficulty concentrating). Indoor air pollutants are inhaled and deposited in the respiratory tract, initially causing inflammation or penetrating through the epithelium into the blood system and, in turn, causing a series of systemic health

A. Kennedy
School of Law, Queensland University of Technology, Brisbane, QLD, Australia

L. Morawska (✉)
School of Earth and Atmospheric Sciences, Queensland University of Technology, Brisbane, QLD, Australia

Global Centre for Clean Air Research (GCARE), School of Sustainability, Civil and Environmental Engineering, Faculty of Engineering and Physical Sciences, University of Surrey, Guildford, UK
e-mail: l.morawska@qut.edu.au

V. Adotey
School of Earth and Atmospheric Sciences, Queensland University of Technology, Brisbane, QLD, Australia

K. M. Kanninen, A. R. White (eds.), *Air Pollution and the Brain*,
https://doi.org/10.1007/978-3-031-99302-2_9

problems. The smallest size fraction of airborne particles, ultrafine particles (UFPs; <100 nm), are of particular importance because, due to their small size, they can penetrate to the deepest parts of the respiratory tract, from where they can reach the blood system.

These particles are emitted in all anthropogenic combustion processes, including the burning of fossil fuels to power vehicles and aircraft, industrial plants, and when cooking or heating with solid fuels or gas, as well as from agricultural burning, forest fires, and other processes in nature. Of particular importance is the less investigated pathway through which UFPs can also cause health problems, which is direct penetration into the brain through the olfactory system. This pathway was already proposed two decades ago [19], and although it has not been systematically investigated due to mainly experimental challenges, its potential impact on the brain and the entire body cannot be overestimated. Given that contemporary people spend more than 90% of their lives breathing indoor air [12] on average 12 times a minute, it is little surprise that poor IAQ ranks as the fifth leading cause of premature deal worldwide [17, 18, 20]. Effective regulation of IAQ is thus critical for public health, workplace safety, and environmental justice. Global approaches to IAQ regulation vary significantly, influenced by cultural, socioeconomic, and environmental factors. While some countries have enacted legislation, most still lack legally mandated performance standards for indoor pollutants. International organisations, including the World Health Organization (WHO) and the United Nations Environment Programme (UNEP), have sought to harmonise approaches to IAQ and provide guidelines for member nations, emphasising the need for equitable access to clean indoor air for vulnerable populations [5]. Despite these advancements, significant challenges remain, including the inconsistent implementation of regulatory approaches, challenges with monitoring and enforcement, and the ongoing threat of emerging pollutants.

This chapter details why there is an urgent need to regulate IAQ more effectively, particularly in addressing the health disparities created by poor air quality within indoor spaces. The chapter reviews international and national approaches to IAQ regulation, examining various governance mechanisms and their effectiveness in protecting public health. Specifically, it explores the differences between guidelines, standards, and binding legislation, as well as the importance of compliance and enforcement measures. Drawing on examples from various jurisdictions, the chapter assesses current strategies, identifying strengths and limitations in existing regulatory frameworks.

9.2 Why Is There a Need for IAQ Regulation?

Public health regulation operates at the nexus of policy, science, and societal values, employing mechanisms such as market oversight, surveillance, and legal mandates to address pressing health issues. These systems aim to mitigate risks, ensure access to essential resources, and promote equitable health outcomes. However, they are not without challenges. Jurisdictional complexities, resource limitations, and rapid technological advancements test the adaptability and efficacy of regulatory frameworks. The COVID-19 pandemic, for instance, underscored the critical need for agile, responsive regulations capable of addressing emerging health threats while maintaining rigorous standards.

Among these challenges, IAQ has emerged as a significant public health concern. Unlike ambient air, IAQ is directly influenced by the built environment and human activity, making it an area ripe for regulatory intervention. With people spending up to 90% of their time indoors, the health impacts of poor IAQ—including respiratory diseases, cardiovascular conditions, and cognitive impairments—position it as an urgent public health issue. Evidence suggests that pollutant concentrations indoors can be higher than outdoor levels due to limited ventilation removing human bio-effluents and pollutant sources such as building materials, cleaning products, and combustion processes [13]. This underscores the critical importance of not only monitoring but also regulating IAQ to mitigate the long-term public health and economic impacts of poor air quality. Particularly concerning is the disproportionate impact on vulnerable groups, such as children, the elderly, pregnant women, and those with pre-existing health conditions. For example, indoor asthma triggers like mould, volatile organic compounds (VOCs), and particulate matter ($PM_{2.5}$) exacerbate health inequalities by disproportionately affecting socio-economically disadvantaged populations who may reside in poorly ventilated or overcrowded environments.

Current global frameworks for managing IAQ rely heavily on non-binding guidelines and voluntary standards, which, while scientifically informed, lack enforceability. The World Health Organization (WHO) has issued IAQ health-based guidelines that recommend thresholds for pollutants such as particulate matter, nitrogen dioxide (NO_2), and carbon monoxide (CO) [22, 23]. However, these guidelines are often adopted inconsistently, if at all, by national and regional authorities [20]. For instance, while some countries incorporate WHO recommendations into their building codes, others lack even basic thresholds for IAQ management. Similarly, voluntary standards developed by technical bodies, such as the American Society of Heating, Refrigerating, and Air-Conditioning Engineers (ASHRAE), provide valuable benchmarks for ventilation and pollutant control [17]. However, without legal frameworks mandating their adoption, these standards remain aspirational, leaving compliance to the discretion of building operators or employers. This creates significant gaps, particularly in residential and public spaces where accountability is diffuse.

9.3 Approaches to Regulation: Distinguishing Between Standards, Guidelines, and Legislation

Addressing IAQ effectively requires careful consideration of the regulatory tools available and how they can be applied to balance enforceability, flexibility, and impact. Regulation serves as a critical mechanism for addressing public health and environmental challenges such as IAQ. However, different purposes and contexts call for different tools, whether voluntary standards, flexible guidelines, or binding legislation. Each of these instruments has its strengths and limitations, and understanding how they can be used to shape behaviour and drive compliance is key to designing effective regulatory frameworks.

Regulation is not merely a matter of prescribing rules; it is a complex process that involves identifying risks, prioritising interventions, and choosing appropriate tools to achieve desired outcomes. Regulatory governance frameworks emphasise that successful regulation must be both strategic and adaptable, capable of responding to diverse challenges across different sectors. Sparrow [21] emphasises that the 'regulatory craft' requires regulators to go beyond setting rules and think critically about how to manage compliance, address risks, and foster accountability.

A central perspective in regulatory theory is the importance of flexibility and proportionality. A graduated approach to enforcement allows regulators to start with softer measures, such as guidelines, before escalating to more coercive mechanisms like binding legislation. This model reflects the notion that regulation should be responsive to the behaviour and motivations of those being regulated, ensuring that interventions are effective and equitable [3]. Moreover, collaborative approaches, where governments, industries, and civil society actors co-design regulatory solutions, offer an effective way to align goals and increase compliance [8].

Standards are a cornerstone of many regulatory frameworks, providing technical specifications or benchmarks to ensure consistency, safety, and quality. They are often developed by independent organisations or industry groups and can either be voluntary or integrated into enforceable systems through adoption by governments or international agreements. Standards are particularly valuable in fostering uniformity across industries, allowing businesses to align practices with shared expectations while driving innovation [11].

A key strength of standards lies in their focus on performance-based outcomes rather than prescriptive methods. For example, ISO 9001, an internationally recognised quality management standard, provides a flexible framework for organisations to improve their processes without dictating how they must do so. This adaptability encourages innovation while maintaining clarity and consistency [11]. Similarly, environmental standards like ISO 14001 have been widely adopted to promote sustainable practices, with research showing tangible improvements in organisational environmental performance following implementation [2].

However, the voluntary nature of standards can also be a limitation, particularly in cases where uptake is not incentivised or where enforcement relies on self-regulation [17]. Standards are often most effective when incorporated into broader regulatory systems, such as public procurement policies or industry accreditation programs, which create additional drivers for compliance. In the context of IAQ, technical standards for ventilation systems, emissions controls, and building materials serve as powerful tools for mitigating risks while allowing flexibility for innovation and cost-effective implementation [22].

Guidelines are another important tool in the regulatory toolkit, offering non-binding recommendations that aim to inform and shape behaviour without the force of law. They are frequently used to address emerging or complex issues where prescriptive rules may be impractical or overly rigid. Unlike standards, which focus on specific benchmarks, guidelines tend to provide practical advice and best practices tailored to different contexts.

There are many different organizations issuing guidelines, and therefore, the basis for the guidelines depends on the focus or priorities of the organisations. A very important class of guidelines is those developed by the WHO, because they are exclusively based on exposure-response relationships found in epidemiological, toxicological, and environment-related studies. No other factors, in particular economic, are considered in developing WHO guidelines but health. The WHO's guidelines on air quality have been instrumental in shaping global responses to air pollution (the recent WHO [24]) and set the direction for indoor air [22].

In general, the adaptability of guidelines makes them particularly well-suited to areas where risks and technologies are constantly evolving. Although these recommendations are not legally binding, they have influenced national policies, encouraged voluntary compliance, and informed the development of technical standards and legislation. In dynamic fields such as public health and environmental management, guidelines offer a critical means of bridging the gap between scientific knowledge and regulatory action.

Despite their advantages, guidelines have limitations. Their non-binding nature can lead to inconsistent application, as adherence often depends on the willingness and capacity of stakeholders to comply. This is particularly challenging in areas like IAQ, where the health impacts of non-compliance can be significant, but enforcement mechanisms may be lacking. Nonetheless, guidelines play a vital role in providing direction and fostering collaboration, often serving as a precursor to more formal regulatory measures [21].

Legislation is the most authoritative and enforceable regulatory tool, establishing binding legal rules that must be followed by individuals and organisations. It provides the legal framework for ensuring compliance and accountability, often specifying penalties for violations and mechanisms for enforcement. In areas where risks are severe or where voluntary compliance is insufficient, legislation becomes essential [4].

One of the key advantages of legislation is its ability to ensure uniform compliance across jurisdictions. For example, in the context of air quality, the *Clean Air Act* in the United States has been credited with significant reductions in pollution levels and associated health impacts. By setting legally enforceable limits on emissions and empowering agencies to monitor and enforce compliance, the legislation has demonstrated the transformative potential of robust regulatory frameworks.

Similarly, mandatory building codes that address IAQ, such as those requiring minimum ventilation standards, can ensure that buildings are designed such that, if operated according to the design, they can mitigate risks and protect public health.

However, legislation also has its limitations. It can be inflexible, particularly in rapidly changing contexts, and the process of enacting new laws or amending existing ones is often slow. This rigidity underscores the importance of integrating legislation with other regulatory instruments, such as standards and guidelines, to create a more adaptive and responsive system. Legislation is most effective when complemented by tools that provide flexibility and foster innovation while maintaining a strong enforcement backbone [8].

The most effective regulatory systems often integrate standards, guidelines, and legislation, leveraging their complementary strengths to address diverse challenges. For example, in addressing IAQ, technical standards can define performance requirements for ventilation systems, guidelines can offer practical advice on implementation, and legislation can establish mandatory compliance and penalties for violations [22]. This layered approach ensures that regulation is both flexible and enforceable, adapting to specific contexts while maintaining accountability.

Theories of regulatory governance provide valuable insights into how these instruments can be combined effectively. A graduated approach to enforcement allows regulators to start with voluntary measures and escalate to binding rules as necessary. This proportionality not only enhances efficiency but also fosters trust and cooperation, aligning regulatory goals with stakeholder interests [3]. Similarly, collaborative frameworks that involve governments, industries, and communities can help ensure that regulatory systems are equitable, inclusive, and grounded in shared expertise [8].

As awareness of the significant health and economic impacts of IAQ grows, regulatory efforts around the world are evolving to address this critical issue. While standards and guidelines have played an essential role in shaping voluntary improvements to date, legislation offers the enforceable framework needed to drive consistent and widespread action. Recent initiatives demonstrate a growing emphasis on creating laws that directly address IAQ concerns in a variety of contexts, from public health to workplace safety and educational environments.

The following section examines legislative efforts aimed at improving IAQ. Beginning with two major US-based proposals, the *Model Clean Indoor Air Act* and the *Indoor Air Quality and Healthy Schools Act*, it explores how policymakers are approaching IAQ through a legislative lens. Following this, it turns to other examples, reviewing legislative frameworks in Belgium, the United Arab Emirates, South Korea, and Taiwan. Together, these examples highlight the diverse strategies being adopted to tackle the complex challenges of regulating IAQ on a global scale.

9.4 Strategies and Pathways for Indoor Air Quality Regulation

As nations increasingly recognise the public health implications of IAQ, there is growing interest in exploring innovative strategies and alternative frameworks to complement existing regulations. Traditional approaches, such as standards, guidelines, and legislation, have laid the groundwork for addressing IAQ issues, yet gaps remain in enforcement, adaptability, and equity. This section investigates novel regulatory options and explores how technological advancements, cross-sectoral collaboration, and public–private partnerships can drive more effective IAQ management. By broadening the scope of interventions and leveraging new tools, governments and industries can overcome current limitations and ensure healthier indoor environments for all.

9.4.1 United States of America

There has been an increased focus on legislative approaches to IAQ in the USA: most notably, the recent the *Model Clean Indoor Air Act* (MCIAA) and the *Indoor Air Quality and Healthy Schools Act* (IAQHSA). Developed by the Johns Hopkins Center for Health Security, the MCIAA is a comprehensive legislative proposal aimed at improving IAQ in public spaces across the United States. To be clear, the MCIAA is not legislation in its own right and does not impose direct standards for IAQ; instead, it proposes a model legal infrastructure that states may utilise to create and enforce regulations. This decentralised approach will allow states to tailor IAQ improvements to their unique conditions while adhering to overarching guidelines. The MCIAA outlines a framework that includes provisions for air quality assessments, mandatory transparency regarding pollutant levels, and incentives for building owners to improve IAQ. By promoting clean building certifications and allowing for tax-deductible expenses related to IAQ enhancements, the Act encourages voluntary compliance among property owners. Key components also involve regulatory oversight by state agencies and the establishment of advisory councils to tailor the implementation process to local needs. Furthermore, the Act proposes to mandate that IAQ testing results be publicly available, fostering accountability and informed decision-making by occupants.

In a similar vein to the *Clean Air Act*, which has successfully reduced outdoor air pollutants that threaten public health, the MCIAA aims to establish guidelines that can lead to improved health outcomes through better management of indoor air contaminants. The Clean Air Act was originally designed to provide uniform federal regulations to ensure a ‘level playing field’ among states, addressing issues related to interstate pollution and maintaining public health standards across the nation. In contrast, the MCIAA focuses on indoor environments such as schools and childcare facilities, aiming to regulate specific pollutants and enhance compliance mechanisms within these spaces.

The MCIAA addresses some of the challenges faced in existing air quality legislation, such as the difficulties of enforcement in diverse settings and the variance in definitions across jurisdictions. By proposing amendments to current public health and environmental protection legislation, the MCIAA encourages states to integrate IAQ standards without the need for entirely new laws.

Building upon the momentum of the MCIAA, the *Indoor Air Quality and Healthy Schools Act* was introduced to Congress in mid-2024 to authorise a national program to reduce the threat to human health posed by exposure to indoor air contaminants [7]. Introduced by US Representatives Brian Fitzpatrick (R-Pa.) and Paul Tonko (D-N.Y.), this bipartisan initiative aims to improve IAQ in educational settings. The proposed legislation addresses long-standing concerns about the health risks associated with poor IAQ in schools and childcare facilities. By focusing on empowering the Environmental Protection Agency (EPA) and offering practical tools for schools, the act seeks to create healthier environments for students and staff.

A key component of the proposed legislation is the expansion of the EPA's Indoor Environments Division. This includes providing enhanced technical assistance to schools and childcare centres to support IAQ improvements. The act also proposes to mandate the establishment of a comprehensive list of significant indoor air contaminants, which will serve as a foundation for targeted mitigation strategies. This approach aims to address pollutants that pose the greatest risks, ensuring schools can prioritise efforts to protect health.

Another major provision of the act is the development of voluntary guidelines for schools to reduce exposure to identified indoor pollutants. While these guidelines would not be legally binding, they will provide practical recommendations for improving IAQ. In addition, the legislation requires a national assessment of IAQ conditions in schools, which will offer valuable data to guide policy decisions and allocate resources effectively.

9.4.2 Belgium

Belgium's Law of 6 November 2022 on the Improvement of Indoor Air Quality in Closed Spaces Accessible to the Public is a landmark piece of legislation aimed at enhancing IAQ in spaces open to the public. The legislation applies to all enclosed spaces accessible to the public, excluding private residences and purely professional areas. It defines key terms such as 'enclosed space', 'open to the public', and 'CO_2 concentration', ensuring clarity in its scope. The law establishes two reference levels for CO_2 concentrations to assess IAQ: Level A, indicating CO_2 concentrations below 900 ppm, and Level B, indicating concentrations below 1200 ppm. These benchmarks are intended to ensure adequate ventilation and air quality in indoor environments, providing operators with measurable targets to maintain healthier conditions.

The law is designed to achieve four overarching objectives:

1. *Defining Reference Levels for Air Quality:* The law establishes indicative, non-mandatory reference levels for good IAQ. While it does not require those responsible for public spaces to ensure high-quality indoor air, these reference levels provide a benchmark for assessing existing air quality conditions.
2. *Encouraging Improvement of IAQ:* The law motivates managers of public spaces to evaluate and improve the air quality within their premises. This is achieved through:

   - The use of CO_2 meters to monitor air quality.
   - Conducting a risk analysis that details the characteristics of the premises, the activities conducted there, potential sources of indoor air pollution, and the air treatment systems in use, such as ventilation and air purification equipment.
   - Developing and implementing an action plan to progressively enhance air quality, potentially through a staged approach.

3. *Promoting Transparency:* To ensure accountability, the law introduces a certification and labelling system. Each public space will be assigned a label that must be displayed prominently. This label will inform the public that the air quality in the space is monitored and controlled, aiming to reduce any adverse health impacts associated with poor air quality.
4. *Expanding Knowledge and Awareness:* The law establishes an Indoor Air Quality Platform dedicated to IAQ, aimed at advancing scientific understanding, advising policymakers, and raising public awareness. This platform fosters collaboration among key stakeholders, including industry representatives, government authorities, ventilation and air purification manufacturers, and the scientific community.

The law provides a phased approach to implementation:

- *From 1 October 2024:* All those responsible for closed spaces accessible to the public can implement the law on a voluntary basis.
- *From 1 January 2027:* The use of at least one air quality meter, development of a risk analysis, and, if necessary, an action plan will become mandatory for certain closed spaces accessible to the public.
- *By the end of 2037:* The scope will be extended to cover all spaces, with the possibility of mandatory certification and labelling requirements coming into force at the earliest on 1 January 2038.

This phased approach allows for gradual adaptation and compliance by stakeholders.

9.4.3 United Arab Emirates

While the UAE has not enacted a singular, overarching IAQ law, it has developed a comprehensive framework that integrates national strategies, technical regulations, and pilot projects. Key elements of this framework include the *National Air Quality Agenda 2031*, which prioritises IAQ alongside outdoor air quality and ambient noise, and the *Cabinet Resolution No. (86) of 2023*, which outlines technical regulations for air quality measurement. Additionally, the UAE has taken practical steps to implement these frameworks through pilot projects, such as the adoption of ASHRAE standards in government buildings. The National Air Quality Agenda 2031, launched by the Ministry of Climate Change and Environment (MOCCAE), provides a general framework to lead and align federal and local government entities, as well as the private sector, to monitor and effectively manage air quality while reducing air pollution. It focuses on four key areas: outdoor air quality, IAQ, ambient odour, and ambient noise. The agenda outlines key enablers, including establishing a clear policy framework, enhancing technical capacities, promoting scientific research, leveraging advanced technology, and facilitating access to financial resources.

The Agenda 2031 document has listed four critical legislative gaps in IAQ in the UAE:

- No comprehensive emission limits for construction materials, furniture, decoration materials, and cleaning products.
- Absence of universally applicable codes of practice for building maintenance activities.
- Lack of a clearly defined strategy and framework for ventilation design requirements.
- Lack of public awareness of IAQ.

Hence, the Agenda 2031 document is founded on three pillars: monitoring, mitigation, and management. The Agenda makes provision for three key IAQ programs and actions to be implemented, but, unlike outdoor air quality, it does not include any numerical values/standards and definite timeframes for the realization of IAQ programs.

In July 2023, the UAE Cabinet issued Resolution No. (86) of 2023 Concerning the Technical Regulations for Air Quality Measurements. This resolution applies to laboratories, environmental monitoring networks, and bodies involved in measuring air pollutants, including IAQ measurements. It mandates that measuring instruments used for air quality assessments are calibrated and possess valid calibration certificates in accordance with national and international standards. Additionally, it requires that air quality measurement laboratories obtain international accreditation in line with the ISO/IEC 17025 standard. The resolution also specifies maximum permissible errors for measuring instruments and outlines procedures for compliance and enforcement.

Finally, the UAE has undertaken pilot projects to implement international IAQ standards. Notably, the Daba Fujairah Court and Prosecution project adopted the American Society of Heating, Refrigerating and Air-Conditioning Engineers (ASHRAE) Standards 241 and 62.1-2022. This initiative resulted in significant improvements in air quality and a 35% reduction in energy use by mechanical ventilation systems. The Ministry of Energy and Infrastructure (MoEI) has expressed its commitment to implementing these standards nationwide.

9.4.4 South Korea

South Korea was the first country in Asia to legislate IAQ with the enactment of the *Air Quality Control in Underground Locations Act* in 1996. This pioneering legislation established comprehensive standards and management practices to safeguard public health by regulating IAQ in underground locations, such as subway stations and underground shopping districts. The most recent iteration, the *Indoor Air Quality Control Act* (IAQ Act), represents a rigorous, health-focused approach to managing indoor air pollution, particularly in public-use facilities and spaces serving vulnerable populations. Unlike countries that rely heavily on voluntary guidelines or broad policy frameworks, such as the United States, South Korea employs a stringent legislative model with enforceable pollutant limits and mandatory monitoring protocols. This reflects a proactive regulatory culture prioritising enforceable standards and transparent accountability to address the risks associated with poor IAQ.

The IAQ Act applies to a wide range of indoor spaces, including general public facilities (e.g., libraries, theatres, underground shopping centres), facilities for sensitive populations (e.g., childcare centres, elderly care facilities), and newly constructed multi-family housing. The *Korean Standards of Indoor Air Quality (KSIAQ)* establishes pollutant limits for spaces under the Act, with thresholds set for particulate matter, nitrogen dioxide, formaldehyde, and total bacteria count, among others.

Key requirements under the Act include:

- Monitoring and Compliance: Regulated facilities must regularly monitor air quality using approved methods and submit reports to oversight authorities. This ensures pollutant levels are measured consistently, and actions are taken to address exceedances.
- Stricter Standards for Sensitive Facilities: Spaces catering to children, the elderly, and other vulnerable populations face more stringent pollutant limits. Studies show these facilities exhibit significantly better air quality due to the implementation of enhanced ventilation and pollutant control systems.
- Transparency and Public Awareness: The Act mandates disclosure of IAQ levels to the public, fostering trust and enabling users to make informed decisions about their indoor environments.

South Korea's Indoor Air Quality Control Act (IAQ Act) has proven effective in improving IAQ, particularly in statutory facilities such as childcare centres and elderly care facilities. Studies comparing statutory and non-statutory facilities indicate that legal regulation has been instrumental in lowering concentrations of key pollutants. For example, a 2019 study demonstrated that statutory facilities recorded significantly lower levels of PM_{10} (46.0 μg/m^3 compared to 49.0 μg/m^3 in non-statutory facilities) and CO_2 (814 ppm compared to 865 ppm) due to the implementation of enhanced ventilation systems and strict compliance measures. Furthermore, none of the statutory facilities exceeded permissible limits for CO, NO_2, or O_3 during the study period, underscoring the impact of monitoring and enforcement [25].

The adaptability of KSIAQ, which underpin the IAQ Act, further enhances its effectiveness. Regular updates to KSIAQ ensure alignment with the latest research. For instance, the inclusion of fine particulate matter ($PM_{2.5}$) and fungi in 2018 reflects growing awareness of indoor air hazards [16, 25].

Despite its successes, the IAQ Act faces notable challenges. A significant limitation is the exclusion of smaller facilities and multi-family housing units from mandatory compliance. Yang et al. [25] found that many non-statutory facilities continue to struggle with high pollutant levels, including formaldehyde (HCHO) and total volatile organic compounds (TVOCs), due to weaker enforcement and inadequate air quality management systems. These gaps leave many populations, especially those in unregulated spaces, vulnerable to poor IAQ.

Regional disparities also hinder the Act's efficacy. Enforcement is stronger in urban centres like Seoul, where resources and technical support are more readily available. However, rural areas often lack the financial and technical capacity to implement and monitor IAQ standards effectively, leading to inconsistent protection across the country [25].

Pollutants, such as microplastics and advanced VOCs, present another challenge. While the KSIAQ already covers a broad spectrum of pollutants, it does not yet address these newer threats, which are increasingly recognised as significant contributors to poor IAQ [16].

Experts in South Korea have recommended several measures to enhance the effectiveness of the Indoor Air Quality Control Act. Expanding the Act's coverage to include smaller public-use facilities and older multi-family housing units is considered critical for addressing existing gaps in protection and ensuring broader public health benefits [25]. Strengthening regional enforcement by allocating additional funding and technical resources to rural and under-resourced areas would help achieve consistent application of IAQ standards nationwide. Experts also emphasise the need to update the KSIAQ to address other classes of pollutants such as microplastics and advanced VOCs, aligning the Act with the latest scientific research and public health priorities [25]. Additionally, promoting sustainable building practices, including the use of eco-friendly materials and energy-efficient ventilation systems, through incentives such as subsidies and tax benefits, is seen as essential for tackling pollution at its source and reducing reliance on reactive measures [25].

9.4.5 Taiwan

Taiwan implemented its *Indoor Air Quality Management Act* in 2012, becoming the second Asian nation to adopt such measures. The Act aims to enhance IAQ and safeguard public health by establishing clear standards and management practices. It mandates that owners, managers, or users of specified public and private premises ensure compliance with established IAQ standards. These premises include educational institutions, medical facilities, government offices, transportation hubs, financial institutions, entertainment venues, and other areas frequented by the public.

Since its enactment, the Indoor Air Quality Management Act has led to increased awareness and proactive management of IAQ in Taiwan. The EPA has conducted training programs for IAQ management personnel and organized public awareness campaigns to promote understanding of IAQ issues. The phased implementation approach allowed facilities time to comply with the new regulations, with initial focus on high-priority locations such as large medical institutions, educational facilities, and transportation hubs.

The Act establishes clear standards and requirements for maintaining IAQ in designated public and private premises across Taiwan. The EPA is tasked with setting and periodically revising IAQ standards to align with emerging scientific research and public health needs. These standards include permissible levels for various pollutants, such as carbon dioxide, carbon monoxide, formaldehyde, TVOCs, bacteria, fungi, particulate matter (PM_{10} and $PM_{2.5}$), and ozone.

Facilities covered under the legislation are required to develop IAQ maintenance and management plans tailored to their specific environments. These plans must designate personnel responsible for ensuring compliance with IAQ standards and implementing proactive measures to maintain air quality. The legislation also emphasises the importance of ongoing assessments, requiring designated premises to conduct regular IAQ monitoring through certified laboratories. Facilities must document the results and make them available for inspection by the EPA to verify compliance with established thresholds.

The Act mandates public disclosure of IAQ conditions within regulated premises. Real-time monitoring equipment must be installed in certain high-priority locations, such as large medical institutions and transportation hubs, with the results prominently displayed at entrances. This approach aims to empower the public with information about the air quality in spaces they frequent, enabling them to make informed decisions about their exposure. The Act also includes robust enforcement mechanisms. The EPA is authorised to conduct on-site inspections, review IAQ records, and impose penalties for non-compliance. Fines for violations range from NT$50,000 to NT$500,000, with the potential for further operational restrictions or business closures in cases of repeated breaches. This enforcement framework ensures that the Act is not merely aspirational but carries tangible consequences for non-adherence, fostering accountability among facility operators.

9.4.6 Assessment of International Approaches

The regulatory approaches to IAQ in the United States, Belgium, the United Arab Emirates (UAE), South Korea, and Taiwan provide a wide-ranging perspective on how different nations have addressed the challenges of regulating IAQ. Each country's approach reflects its public health priorities, governance structures, and resources, leading to distinct strengths and limitations.

South Korea and Taiwan have implemented some of the most comprehensive IAQ frameworks in the world. Both nations rely on enforceable legislation, underpinned by regularly updated pollutant thresholds and robust enforcement mechanisms. South Korea's IAQ Act was the first such law in Asia, setting a precedent for rigorous IAQ regulation. It mandates pollutant limits for PM_{10}, CO_2, NO_2, and VOCs in public-use facilities, with stricter thresholds for spaces catering to vulnerable populations, such as childcare centres and elderly care facilities. Taiwan's Indoor Air Quality Management Act covers a similarly broad range of pollutants while requiring real-time air quality monitoring and public disclosure in high-priority facilities such as schools, hospitals, and transportation hubs. Both models demonstrate measurable effectiveness. Taiwan's enforcement framework further strengthens compliance by imposing substantial penalties for violations and requiring facilities to appoint dedicated IAQ management personnel. This focus on accountability ensures adherence to pollutant thresholds, fostering healthier indoor environments.

The success of these legislative approaches underscores the importance of binding regulations in achieving consistent outcomes. By mandating pollutant limits, monitoring, and enforcement, South Korea and Taiwan provide an example for nations seeking to institutionalise IAQ standards. However, their frameworks are not without challenges, including resource disparities and the need to address other classes of pollutants.

In contrast to South Korea and Taiwan's legislative models, Belgium and the UAE employ more flexible frameworks, emphasising voluntary compliance, strategic initiatives, and public awareness. While these thresholds are not legally enforceable, they serve as benchmarks for assessing and improving IAQ. A notable innovation in Belgium's approach is its certification and labelling system, which requires public-use facilities to display air quality labels, informing occupants of IAQ conditions and fostering transparency. Belgium's model reflects a phased regulatory approach, balancing immediate public health benefits with long-term regulatory certainty. Rather than imposing strict compliance measures from the outset, the framework allows for gradual adaptation by stakeholders, ensuring that businesses, building owners, and facility managers have time to adjust to new expectations. By 2037, the law is set to expand to cover all spaces, and from 1 January 2038, mandatory certification and labelling requirements may come into force. This staged approach illustrates a regulatory design strategy used to build compliance capacity, encourage voluntary improvements, and allow for iterative policy adjustments based on data and experience. It reflects a broader trend in environmental and public health regulation, where governments seek to create a regulatory pathway that encourages early adoption and industry cooperation before enforcing mandatory standards. Meanwhile, the UAE's National Air Quality Agenda 2031 takes a strategic and integrated approach, aligning IAQ initiatives with broader environmental and sustainability goals. While the UAE does not have a comprehensive, standalone IAQ law, its regulatory framework leverages pilot projects, such as adopting ASHRAE standards in government buildings, and establishes technical regulations for air quality monitoring and assessment. Both Belgium and the UAE illustrate how innovation and stakeholder engagement can drive IAQ improvements, albeit through different mechanisms. Belgium's certification and labelling system enhances public awareness and empowers individuals to make informed decisions about indoor environments, while the UAE's pilot projects serve as a testing ground for scalable regulatory and technological solutions. However, the absence of binding, enforceable IAQ thresholds in both frameworks (at least in the short term, in the case of Belgium) presents a regulatory gap that may limit long-term effectiveness, particularly in facilities that lack the incentives or resources to voluntarily improve IAQ.

The United States presents a unique case, with its reliance on state-level implementation of model legislation. Such a decentralised model will enable tailored solutions that reflect local priorities and capacities. However, it may also introduce variability in implementation, with some states potentially adopting rigorous IAQ legislation while others lag behind. Such inconsistency will pose challenges for ensuring equitable health outcomes, particularly in under-resourced regions where IAQ issues may be most acute. The reliance on voluntary adoption further underscores the need for federal leadership to standardise IAQ benchmarks and promote uniform enforcement.

A common thread across all jurisdictions is the need to expand the focus to include more pollutants and adapt to new health risks. Traditional IAQ standards often focus on pollutants like CO_2, $PM_{2.5}$, and formaldehyde, but newer contaminants such as microplastics, advanced VOCs, and airborne pathogens require updated regulatory frameworks. Taiwan and South Korea have taken steps to expand their pollutant lists, but gaps remain in addressing the full spectrum of indoor air hazards.

Technological advancements present an opportunity to tackle these challenges. Low-cost sensors, IoT-enabled monitoring systems, and data analytics can enhance real-time IAQ assessments and improve compliance. For example, Taiwan's requirement for automatic monitoring equipment in high-priority facilities demonstrates how technology can support transparency and accountability [6]. Scaling such innovations across jurisdictions could provide a cost-effective way to enforce standards and identify emerging risks.

Another critical consideration is the equitable distribution of IAQ protections. While South Korea and Taiwan lead in regulatory rigor, rural areas and non-statutory facilities often lack the resources to comply with IAQ standards. Similarly, the United States' decentralised model risks exacerbating disparities, with wealthier states implementing stronger protections than their less-resourced counterparts.

Table 9.1 Type of regulations enacted by different counties

Country	Type of regulation/ name year	Pollutant	Realtime or periodic	Anticipated regulatory impact
Taiwan	IAQ Management Act (Government regulation-2012)	Carbon dioxide (CO_2), carbon monoxide (CO), formaldehyde, ozone, PM_{10}, $PM_{2.5}$, TVOCs, bacteria, fungi	Periodic	Expected reduction in IAQ pollutant exposure levels.
South Korea	IAQ control Act (Government regulation-2015)	PM_{10}, $PM_{2.5}$, radon, NO_2, formaldehyde, CO_2, CO	Periodic	Reduction in concentration levels of indoor pollutants, improving health outcomes.
UAE	National Air Quality Agenda 2031 (Government guidelines-2022)	Carbon dioxide (CO_2), carbon monoxide (CO), formaldehyde, ozone, PM_{10}, $PM_{2.5}$	Not yet determined	Expected positive health, environmental, and economic impacts; details pending further research.
Belgium	Clean Air Law (Government regulation-2022)	Carbon dioxide (CO_2)	Periodic	Projected reduction in indoor CO_2 levels by 2038.
USA	Model Clean Indoor Air Act (Government model legislation-2023)	Carbon dioxide (CO_2), carbon monoxide (CO), formaldehyde, ozone, PM_{10}, $PM_{2.5}$	To be determined by individual States	Expected to reduce indoor air pollution, with effectiveness varying by state implementation.

Addressing these inequities requires targeted interventions. Subsidies, technical support, and capacity-building programs can help under-resourced facilities meet IAQ requirements. Public awareness campaigns, as seen in Belgium and the UAE, also play a vital role in promoting voluntary compliance and empowering communities to advocate for better IAQ (Table 9.1).

9.4.7 *Utilising Existing Regulatory Frameworks*

One of the most effective strategies for advancing IAQ regulation is leveraging existing legislative frameworks, such as work health and safety (WHS) laws, environmental protection acts, and building codes. These frameworks provide established structures for enforcement and monitoring, reducing the need for entirely new legislative mechanisms. By integrating IAQ standards into these systems, governments can address workplace exposures, reduce compliance costs for businesses, and streamline oversight.

Work health and safety laws are particularly relevant because poor IAQ is a recognised occupational hazard. Incorporating IAQ standards into WHS regulations ensures that employers are legally obligated to maintain healthy air environments for their employees. Taiwan's Indoor Air Quality Management Act, for instance, requires workplace monitoring in certain facilities, tying IAQ assessments to broader occupational safety mandates [6]. Similarly, South Korea's Indoor Air Quality Control Act aligns with workplace safety objectives by targeting high-risk environments, such as underground facilities and industrial spaces, which are particularly prone to poor ventilation and pollutant build-up [25]. Such integration allows IAQ regulation to benefit from established enforcement mechanisms, including workplace inspections and hazard assessments, creating synergies between IAQ management and occupational health.

Belgium also offers a promising model for utilising existing frameworks. The Law of 6 November 2022 requires premises to conduct risk analyses and implement action plans for improving IAQ, aligning closely with the risk assessment protocols already mandated under European workplace safety laws [9]. This approach reduces the administrative burden for businesses, as it builds on processes they are already required to undertake for other health and safety purposes.

The integration of IAQ into WHS laws is not without challenges. It requires a clear delineation of responsibilities, ensuring that IAQ is prioritised alongside other occupational hazards. Additionally, enforcement agencies must be adequately resourced to handle the increased scope of inspections and compliance checks. Nonetheless, leveraging existing WHS frameworks provides a practical and cost-effective pathway to improve IAQ protections while reinforcing the importance of workplace safety.

9.5 Conclusion

The regulation of IAQ is a pressing global issue with significant implications for public health, environmental justice, and economic development. This article has highlighted the strengths and limitations of various regulatory approaches, from the flexibility of voluntary standards to the enforceability of binding legislation. Case studies from countries such as the United States, Belgium, the United Arab Emirates, South Korea, and Taiwan reveal both the progress made and the challenges that persist in ensuring consistent IAQ management across diverse contexts.

To move forward effectively, policymakers must adopt a holistic and integrated approach, leveraging the complementary strengths of standards, guidelines, and legislation. Emphasising collaboration among governments, industries, and civil society is crucial to fostering compliance, innovation, and equity. Additionally, adapting to emerging pollutants and incorporating advancements in monitoring technology will be essential for building resilient IAQ frameworks. As awareness grows and frameworks evolve, the goal must remain clear: creating indoor environments that prioritise health, sustainability, and inclusivity for all.

Acknowledgement This work is supported by the Australia Research Council (ARC) Industrial Transformation Training Centres (ITTC) 'ARC Training Centre for Advanced Building Systems Against Airborne Infection Transmission (THRIVE)' (IC220100012).

References

1. Asikainen, A., et al. (2016). Reducing burden of disease from residential indoor air exposures in Europe (HEALTHVENT project). *Environmental Health, 15*, 61–72.
2. Boiral, O., & Henri, J.-F. (2012). Modelling the impact of ISO 14001 on environmental performance: A comparative approach. *Journal of Environmental Management, 99*, 84–97.
3. Braithwaite, J. (2008). *Regulatory capitalism: How it works, ideas for making it work better*. Edward Elgar Publishing.
4. Dockery, D. W., Pope, C. A., Xu, X., Spengler, J. D., Ware, J. H., Fay, M. E., Ferris, B. G., Jr., & Speizer, F. E. (1993). An association between air pollution and mortality in six US cities. *New England Journal of Medicine, 329*, 1753–1759.
5. Emmerich, S., Toyinbo, O., Hagerhed, L., Dimitroulopoulou, S., Dudzinska, M., Hemming, D., Park, J.-H., & Haverinen-Shaughnessy, U. (2022). Open database for international and national indoor environmental quality guidelines. *Indoor Air [Online]*. https://doi.org/10.1111/ina.13028
6. EPA Taiwan. (2011). *Indoor air quality act in: (EPA) E.P.A., ed. Laws & regulations databse of the Republic of China (Taiwan)*. https://law.moj.gov.tw/ENG/LawClass/LawAll.aspx?pcode=O0130001
7. GovTrack US. (2024). *H.R. 9131 (118th): Indoor air quality and healthy schools act of 2024*. https://www.govtrack.us/congress/bills/118/hr9131
8. Gunningham, N., Grabosky, P., & Sinclair, D. (1998). *Smart regulation: Designing environmental policy*. Oxford University Press.
9. Health Belgium. (2022). *Legal framework regarding indoor air quality*. FPS Public Health. https://www.health.belgium.be/en/closer-legal-framework-indoor-air-quality
10. Hänninen, O., et al. (2014). Environmental burden of disease in Europe: Assessing nine risk factors in six countries. *Environmental Health Perspectives, 122*, 439–446.
11. International Organization for Standardization [ISO]. (2015). *ISO 9001: Quality management systems—Requirements*. Retrieved from https://www.iso.org
12. Klepeis, N. E., et al. (2001). The National Human Activity Pattern Survey (NHAPS): A resource for assessing exposure to environmental pollutants. *Journal of Exposure Science & Environmental Epidemiology, 11*, 231–252.
13. Leung, D. Y. (2015). Outdoor-indoor air pollution in urban environment: Challenges and opportunity. *Frontiers in Environmental Science, 2*, 69.
14. Liu, N., et al. (2023). The burden of disease attributable to indoor air pollutants in China from 2000 to 2017. *Lancet Planet Health, 7*, e900–e911.
15. Logue, J. M., Price, P. N., Sherman, M. H., & Singer, B. C. (2012). A method to estimate the chronic health impact of air pollutants in U.S. residences. *Environmental Health Perspectives, 120*, 216–222.
16. Ministry of Environment. (2021). *Indoor air quality standards and management policies*. Korea R.O. https://me.go.kr
17. Morawska, L., Allen, J., Bahnfleth, W., Bennett, B., Bluyssen, P. M., Boerstra, A., Buonanno, G., Cao, J., Dancer, S. J., & Floto, A. (2024). Mandating indoor air quality for public buildings. *Science, 383*, 1418–1420.
18. Morawska, L., & Huang, W. (2022). *Chapter X.I: WHO health guidelines for indoor air quality and national recommendations/standards*. https://link.springer.com/referenceworkentry/10.1007/978-981-10-5155-5_49-1

19. Oberdörster, G., Sharp, Z., Atudorei, V., Elder, A., Gelein, R., Kreyling, W., & Cox, C. (2004). Translocation of inhaled ultrafine particles to the brain. *Inhalation Toxicology, 16*(6–7), 437–445. https://doi.org/10.1080/08958370490439597. PMID: 15204759.
20. Siddique, A., Al-Shamlan, M. Y., Al-Romaihi, H. E., & Khwaja, H. A. (2023). Beyond the outdoors: Indoor air quality guidelines and standards–challenges, inequalities, and the path forward. *Reviews on Environmental Health, 40*(1), 21–35.
21. Sparrow, M. K. (2011). *The regulatory craft: Controlling risks, solving problems, and managing compliance*. Rowman & Littlefield.
22. WHO. (2010). *Guidelines for indoor air quality, selected pollutants*. WHO, Geneva. https://www.euro.who.int/data/assets/pdf_file/0009/128169/e94535.pdf
23. WHO. (2021). *WHO global air quality guidelines: Particulate matter (PM2.5 and PM10), ozone, nitrogen dioxide, sulfur dioxide and carbon monoxide*. World Health Organization. https://apps.who.int/iris/handle/10665/345329
24. World Health Organization. (2021). *WHO global air quality guidelines. Who guidelines for indoor air quality: Selected pollutants*. WHO, 2010. https://www.who.int/publications/i/item/9789240034228?ua=1
25. Yang, J., Seo, J.-H., Jeong, N.-N., & Sohn, J.-R. (2019). Effects of legal regulation on indoor air quality in facilities for sensitive populations—A field study in Seoul, Korea. *Environmental Management, 64*, 344–352.

Index

K. M. Kanninen, A. R. White (eds.), *Air Pollution and the Brain*,
https://doi.org/10.1007/978-3-031-99302-2

P

R

S

T

U

V

W

GPSR Compliance

The European Union's (EU) General Product Safety Regulation (GPSR) is a set of rules that requires consumer products to be safe and our obligations to ensure this.

If you have any concerns about our products, you can contact us on ProductSafety@springernature.com

In case Publisher is established outside the EU, the EU authorized representative is:

Springer Nature Customer Service Center GmbH
Europaplatz 3
69115 Heidelberg, Germany

Batch number: 10755332

Printed by Printforce, the Netherlands